HANDBOOK OF

Mathematical Techniques for Wave/Structure Interactions

HANDBOOK OF
Mathematical Techniques for Wave/Structure Interactions

C. M. Linton
P. McIver

CHAPMAN & HALL/CRC

Boca Raton   London   New York   Washington, D.C.

Library of Congress Cataloging-in-Publication Data

Linton, Christopher M.
    Handbook of mathematical techniques for wave/structure interactions
/ Christopher M. Linton, Philip McIver.
        p. cm.
    Includes bibliographical references and index.
    ISBN 1-58488-132-1 (alk. paper)
    1. Water waves--Mathematical models--Handbooks, manuals, etc. 2.
Structural dynamics--Handbooks, manuals, etc. I. McIver, Philip.
II. Title.
TA654.55 .L56 2001
624.1′71--dc21

                                             2001017068
                                                CIP

Visit the CRC Press Web site at www.crcpress.com

Contents

Appendices

Preface

A wide range of mathematical techniques are now available for the solution of problems involving the interaction of waves with structures. Many of these techniques are described in existing textbooks, but often not in the context of wave/structure interactions and often without reference to applications at all. This book draws together some of the most important of these methods into a single text to form a convenient reference work for both applied mathematicians and engineers. All of the techniques are described within the context of wave/structure interactions and are often illustrated by application to research problems. An advantage of describing a number of methods within the same text is that, for particular problems, direct comparisons can be made between them.

The methods described in this book may be applied to a wide variety of problems from many fields of research including water waves, acoustics, electromagnetic waves, waves in elastic media, and solid-state physics. When writing the book it soon became clear that it was impossible to do justice to all of these fields, and so we decided to focus mainly on problems that have interpretations within the linearized theory of water waves. However, we have made extensive reference to applications of the techniques in other areas, both throughout the text and in extensive bibliographical notes that are placed at the end of many of the sections within the book. Our hope is that in this way the book will be a useful reference work for workers from a wide range of research fields.

The reader is assumed to have a knowledge at an undergraduate level of multivariable calculus, including the solution of linear partial differential equations, and complex-variable theory. Detailed explanations are given of the important steps within the mathematical development and, where possible, physical interpretations of mathematical results are given. The overall aim is provide a pedagogical text that will help readers apply the techniques to their own problems.

In view of our decision to focus on water waves, the first chapter of the book is concerned with the linearized theory of the interaction of water waves and structures. One purpose of this chapter is to help those unfamiliar with the area to appreciate more the applications discussed in later chapters. We also took the view that it would be convenient for those who work in water waves to have a fairly detailed account of background material readily to hand.

Each subsequent chapter deals with a different technique. Chapters 2 and 3 consider the representation of solutions by infinite series of suitably chosen functions. Chapter 2 first describes the construction of eigenfunction expansions using the technique of separation of variables for partial differential equations, and then goes on to describe the method of matched eigenfunction expansions. Chapter 3 deals with so-called multipole expansions where the basis functions are singular at a point exterior to the domain of interest. In some simple cases the multipoles are exactly the eigenfunctions obtained by separation of variables.

Many problems are conveniently formulated in terms of integral equations and in Chapter 4 some methods for the formulation and numerical solution of such equations are described. In the water-wave problem numerical methods based on integral equations obtained from an application of Green's theorem are popular. Crucial to the success of these methods is the efficient evaluation of the Green's function and Chapter 4 also describes effective methods for this purpose.

The Wiener-Hopf method and the related residue calculus theory are important techniques for the solution of problems of wave interaction with semi-infinite geometries. A detailed exposition of these advanced techniques is given in Chapter 5. Examples are used to illustrate how both techniques can be used to obtain approximations to the solutions of problems involving finite geometries.

Chapter 6 deals with wave interaction with an array of structures and describes both exact methods and very effective approximate techniques based on the assumption that the structures are widely spaced compared to the wavelength. Arrays that are compact or that extend to infinity in one direction are both treated.

It is often useful to use approximate techniques on difficult problems in order to gain insight into the physical processes involved. It is with this in mind that approximate methods are given in Chapter 7 for the analysis of wave interaction with objects that are small relative to the wavelength. In this chapter, the method of matched asymptotic expansions is applied to both small structures and structures with small gaps. A separate technique for the solution of eigenvalue problems involving small objects is also described.

Finally, Chapter 8 describes variational techniques that can, for certain classes of problems, yield very accurate solutions with minimal computational effort. First of all, it is shown how variational methods can be used to improve the eigenfunction techniques given in Chapter 2, and then the classical Rayleigh-Ritz method for the solution of eigenvalue problems is described.

Almost all of the numerical computations reported here were carried out by the authors specifically for inclusion in this book. We also believe that some of the material in this book appears in print for the first time. This includes the multipole solution for oblique-wave incidence on a submerged cylinder in §3.1.1, the development of the wide-spacing approximation for scattering and radiation by an arbitrary number of structures described in §6.3, and the higher-order solution to the breakwater-gap problem by matched asymptotic expansions given in §7.2.3

We are very grateful to Paul Martin and David Porter for answering our technical queries, and to Maureen McIver for her critical reading of the manuscript. Thanks are also due to John Cadby for some of the work in §3.1.1 and to Matthew Bowen for working through some sections and weeding out mistakes. Some must remain, and for that we apologise.

C. M. Linton
P. McIver

Chapter 1

The water-wave problem

The techniques described in this book have applications in many areas of physical interest, including the fields of water waves, acoustics, and electromagnetic waves. However, for definiteness in the interpretation of both the problems and their solutions, we have decided to focus most attention on applications in the linearized theory of water waves. Examples of the application of the techniques to problems in other physical contexts are referred to, where appropriate, in the text and are also provided through the extensive bibliographical notes. For the scattering of electromagnetic and acoustic waves by structures, an extensive collection of results is given by Bowman, Senior, and Uslenghi (1987).

The present introductory chapter is therefore concerned with the hydrodynamics of the interaction of water waves with structures. It includes the equations that govern the fluid motion as well as definitions of integrated quantities which are physically important, such as hydrodynamic forces and reflection and transmission coefficients.

In §1.1, we begin with brief derivations of the governing equations and nonlinear boundary conditions for the water-wave problem; the reduction to the linearized theory for small amplitude waves is described in §1.2. When subject to an incident water wave, a structure will in general scatter the incident wave field, and be forced to move so that further waves are radiated. On the basis of the linearized theory, the full problem can be decoupled into a scattering problem and a radiation problem which are linked by the equation of motion for the structure. In §1.3 the scattering and radiation problems are described in detail, and expressions are given for the hydrodynamic forces on a structure in both problems. Some identities involving these forces, as well as other integrated quantities, are discussed in §1.4. Finally, in §1.5, the relationship between the hydrodynamic forces in the radiation problem and the energy of the fluid motion is described.

1.1   Introduction

Here a brief introduction is given to the standard equations of inviscid water-wave theory. For a more complete account the reader might consult Chapter 1 of the book by Crapper (1984) or of that by Johnson (1997).

For all of the water-wave problems discussed in this book, a Cartesian coordinate system (x, y, z) is adopted with the z-axis directed vertically upwards and with $z = 0$ in the plane of the undisturbed free surface. For purely two-dimensional motion the dependence on y will be omitted and, throughout, time is denoted by t.

The fluid is assumed to be inviscid and incompressible and its motion to be irrotational. For irrotational motion the fluid velocity $\mathbf{u}$ may be expressed as the gradient of a scalar velocity potential $\Phi(x, y, z, t)$, that is $\mathbf{u} = \nabla\Phi$. Conservation of mass requires that the divergence of the velocity is zero so that Φ satisfies Laplace's equation

$$\nabla^2 \Phi = 0 \tag{1.1}$$

throughout the fluid.

The vertical elevation of a point on the free surface is written

$$z = \eta(x, y, t). \tag{1.2}$$

The kinematic condition that fluid particles cannot cross the air-water interface is obtained by equating the vertical speed of the free surface itself to that of a fluid particle in the free surface to get

$$\frac{\partial\eta}{\partial t} + \frac{\partial\Phi}{\partial x}\frac{\partial\eta}{\partial x} + \frac{\partial\Phi}{\partial y}\frac{\partial\eta}{\partial y} = \frac{\partial\Phi}{\partial z} \quad \text{on} \quad z = \eta(x, y, t). \tag{1.3}$$

If surface tension is neglected (this is valid for waves longer than a few centimetres), the pressure must be continuous across the interface, and at any point in the fluid Bernoulli's equation,

$$\frac{\partial\Phi}{\partial t} + \tfrac{1}{2}|\nabla\Phi|^2 + \frac{p}{\rho} + gz = 0, \tag{1.4}$$

holds where ρ is the fluid density, g is the acceleration due to gravity and p is the pressure in the fluid relative to atmospheric pressure. Because of the comparatively small density of the air its motion may be neglected and the pressure along the interface taken to be constant. Bernoulli's equation evaluated at the interface, where $p = 0$, then gives the dynamic condition

$$\frac{\partial \Phi}{\partial t} + \tfrac{1}{2}|\nabla \Phi|^2 + g\eta = 0 \quad \text{on} \quad z = \eta(x, y, t). \tag{1.5}$$

When there is an impermeable sea bed so that the local fluid depth is $h(x, y)$, then there must be no flow normal to the bed and hence

$$\frac{\partial \Phi}{\partial n} = 0 \quad \text{on} \quad z = -h(x, y), \tag{1.6}$$

where n is a coordinate measured normal to the bed.

1.2   The linearized equations

For sufficiently small motions relative to the wavelength, the above nonlinear free-surface conditions (1.3) and (1.5) may be linearized about the undisturbed state. The linearized theory requires the amplitude of the fluid motion to be small compared to the wavelength throughout the fluid domain including the vicinity of any structures, and hence the amplitude of any structural motions must also be small relative to other length scales. It is consistent with the linearization to apply the free-surface boundary conditions on $z = 0$, in which case the kinematic condition (1.3) becomes

$$\frac{\partial \eta}{\partial t} = \frac{\partial \Phi}{\partial z} \quad \text{on} \quad z = 0 \tag{1.7}$$

and the dynamic condition (1.5) becomes

$$\frac{\partial \Phi}{\partial t} + g\eta = 0 \quad \text{on} \quad z = 0. \tag{1.8}$$

These two conditions may be combined by differentiation of (1.8) with respect to t and substitution for $\partial \eta / \partial t$ from (1.7) to get the linearized free-surface condition

$$\frac{\partial^2 \Phi}{\partial t^2} + g\frac{\partial \Phi}{\partial z} = 0 \quad \text{on} \quad z = 0. \tag{1.9}$$

All of the results, here and in subsequent chapters, are based on the linearized theory and relate to time-harmonic motion with a specified frequency. Corresponding results for more complicated motions such as the response to a random sea may be constructed by Fourier analysis (see for example Faltinsen 1990, Chapter 2). For time-harmonic motions of radian frequency ω, time may be removed from the problem by writing

$$\Phi(x, y, z, t) = \text{Re}\left\{\phi(x, y, z)\, \text{e}^{-\text{i}\omega t}\right\}, \tag{1.10}$$

where ϕ is a complex-valued potential and Re denotes the real part. This is a convenient way of extracting the time dependence in time-harmonic problems as only a single function need be solved for. A more cumbersome alternative is to write

$$\Phi(x, y, z, t) = \phi_{\text{r}}(x, y, z) \cos \omega t + \phi_{\text{i}}(x, y, z) \sin \omega t, \tag{1.11}$$

where ϕ_{r} and ϕ_{i} are the real and imaginary parts of ϕ, and then to solve for the two real-valued functions ϕ_{r} and ϕ_{i} separately. A comparison of (1.10) and (1.11) shows that ϕ contains information about both the amplitude and the phase of the motion.

From equation (1.1) it follows that ϕ satisfies Laplace's equation

$$\nabla^2 \phi = 0 \tag{1.12}$$

throughout the fluid domain. In terms of ϕ, the linearized free-surface condition (1.9) becomes

$$\frac{\partial \phi}{\partial z} = K\phi \quad \text{on} \quad z = 0, \tag{1.13}$$

where $K = \omega^2/g$, and the bed condition (1.6) becomes

$$\frac{\partial \phi}{\partial n} = 0 \quad \text{on} \quad z = -h(x, y). \tag{1.14}$$

For deep water, characterized by the limit $h(x, y) \to \infty$ for all x and y, the condition (1.14) is replaced by

$$|\nabla \phi| \to 0 \quad \text{as} \quad z \to -\infty. \tag{1.15}$$

The potential ϕ must also satisfy boundary conditions on any structures within the fluid and radiation conditions at large horizontal distances; these will be discussed in §§1.3.1–1.3.3.

We will only consider the cases of deep water and constant finite depth in any detail. A great deal of work has been done on scattering problems in which h is variable; for a recent review of two-dimensional problems of this type, see Porter and Chamberlain (1997). In the case of constant finite depth h, one solution of (1.12)–(1.14) is a plane wave. With horizontal polar coordinates (r, θ) defined by

$$x = r \cos \theta, \quad y = r \sin \theta, \tag{1.16}$$

a plane wave of amplitude A and wavenumber k propagating in the direction $\theta = \beta$ has a potential

$$\phi_{\text{I}} = -\frac{igA}{\omega \cosh kh} \, \mathrm{e}^{ikr \cos(\theta - \beta)} \cosh k(z + h), \tag{1.17}$$

corresponding to a free-surface elevation given, from (1.8) and (1.10), by

$$\eta = A \, \mathrm{e}^{ikr \cos(\theta - \beta)} . \tag{1.18}$$

The wavenumber k is the number of wavelengths in a distance 2π, measured in the direction of wave propagation, and so the wavelength of such a disturbance is $\lambda = 2\pi/k$. If the wave is propagating in the direction of $\beta = 0$ along the x-axis, then the exponential factor in (1.17) and (1.18) reduces to e^{ikx}. The potential ϕ_{I} identically satisfies (1.12) and (1.14) and will satisfy (1.13) provided $K = \omega^2/g$ and k are related by the dispersion relation

$$K = k \tanh kh. \tag{1.19}$$

The solutions to this equation will be discussed in detail in §2.1. Here we just note that for a specified frequency ω the equation determines the wavenumber k.

In the case of deep water the dispersion relation reduces to $K = k$ and a plane wave making an angle $\beta = 0$ with the positive x axis has the form

$$\phi_{\text{I}} = -\frac{igA}{\omega} \, \mathrm{e}^{iKr \cos(\theta - \beta)} \, \mathrm{e}^{Kz}, \qquad \eta = A \, \mathrm{e}^{iKr \cos(\theta - \beta)} . \tag{1.20}$$

1.3 Interaction of a wave with a structure

On the surface of a structure, the normal component of the structural velocity must equal the velocity component in the same direction of an adjacent fluid particle. In terms of the velocity potential Φ introduced in §1.1, this requires

$$\frac{\partial \Phi}{\partial n} = V_n, \tag{1.21}$$

where V_n is the component of the structural velocity in the direction of the normal coordinate n directed out of the fluid. In the linearized theory this condition is applied on the equilibrium surface of the structure which will be denoted by S_B.

A wave train incident upon a floating structure will be diffracted to produce a scattered wave field and also set the structure in motion to produce a radiated field. By linear superposition, the velocity potential may be decomposed into two parts as

$$\Phi = \Phi_S + \Phi_R. \tag{1.22}$$

The potential Φ_S is the solution of the scattering problem in which the structure is held fixed in the waves and it may be further decomposed as

$$\Phi_S = \Phi_I + \Phi_D \tag{1.23}$$

where Φ_I represents the incident wave train and Φ_D the diffracted waves. Because the structure is held fixed, the appropriate boundary condition is

$$\frac{\partial \Phi_S}{\partial n} = 0 \quad \text{or} \quad \frac{\partial \Phi_D}{\partial n} = -\frac{\partial \Phi_I}{\partial n} \quad \text{on} \quad S_B. \tag{1.24}$$

The potential Φ_R is the solution of the radiation problem, in which the structure is forced to oscillate in the absence of an incident wave, and satisfies

$$\frac{\partial \Phi_R}{\partial n} = V_n \quad \text{on} \quad S_B. \tag{1.25}$$

In general, the normal velocity V_n is found from the equation of motion of the structure (see Mei 1983, §7.2, or Newman 1977, Chapter 6) and will depend, in particular, on the forces that result from any incident waves.

Following (1.10), for time-harmonic motions with radian frequency ω, the time variation in the scattering potential is separated out by writing

$$\Phi_S = \mathrm{Re}\left\{(\phi_I + \phi_D)\,e^{-i\omega t}\right\}, \tag{1.26}$$

where, for constant finite depth, the incident wave ϕ_I has the form (1.17) (or (1.20) for deep water). Similarly, for the radiation potential the time variation is separated out by writing

$$\Phi_R = \text{Re}\left\{\phi_R\, e^{-i\omega t}\right\}. \tag{1.27}$$

Both ϕ_D and ϕ_R are complex-valued functions of position only.

1.3.1 The radiation condition

To obtain a unique solution, the diffracted field ϕ_D and the radiated field ϕ_R, defined in equations (1.26) and (1.27), respectively, must each satisfy a radiation condition specifying that the waves corresponding to these potentials propagate away from the structure. For w equal to either ϕ_D or ϕ_R the radiation condition can be written

$$\lim_{kx\to\pm\infty}\left(\frac{\partial w}{\partial x} \mp ikw\right) = 0 \tag{1.28}$$

in two dimensions and

$$\lim_{kr\to\infty} r^{1/2}\left(\frac{\partial w}{\partial r} - ikw\right) = 0 \tag{1.29}$$

in three dimensions, where k is the wavenumber introduced in §1.2 and r is the horizontal polar coordinate introduced in equation (1.16). In three dimensions a radially spreading cylindrical wave of decreasing amplitude is obtained and energy conservation arguments require the factor of $r^{1/2}$ in (1.29). The role of the precise form of the radiation condition (1.29) in formulating integral equations is demonstrated in §4.2.1.

Here it is assumed that there are no 'trapped-mode' solutions at the frequency of interest. A trapped mode supported by the structure is a free oscillation of the fluid that has finite energy. The existence of such a mode would mean that the radiation condition no longer guarantees the uniqueness of the solution. Uniqueness has been proved for certain classes of geometries; for example, uniqueness of the solution is established for surface-piercing structures which have the property that any vertical line emanating from the free surface does not intersect the body (John 1950), or two-dimensional structures which are contained within lines which emanate from the free surface at a certain angle (Simon and Ursell 1984). On the other hand, trapped modes have been proved to exist for certain surface-piercing structures in the two-dimensional water-wave problem by McIver (1996a) and in the three-dimensional problem by McIver and McIver (1997). Their existence for certain submerged structures has been demonstrated by McIver (2000a). The implications of the existence of trapped modes for the solutions of the boundary-value problem are discussed by McIver (1997), while the implications for the hydrodynamic forces, defined later in §1.3.4, are discussed by Newman (1999).

1.3.2   The scattering problem

In the scattering problem, the solution ϕ_D for the diffracted wave field, defined in equation (1.26), is a solution of Laplace's equation within the fluid domain and satisfies the free-surface condition (1.13), the bed condition (1.14) (or (1.15) for deep water), and

$$\frac{\partial \phi_D}{\partial n} = -\frac{\partial \phi_I}{\partial n} \quad \text{on} \quad S_B. \tag{1.30}$$

As noted in §1.3.1, to ensure a unique solution the diffracted field must also satisfy a radiation condition in the form of either (1.28) or (1.29) with $w = \phi_D$.

1.3.3   The radiation problem

To fully describe the position and orientation of a rigid structure, six coordinates, corresponding to six modes of motion, are required. In naval hydrodynamics, the translational modes are called surge, sway, and heave for the motions parallel to the x-, y- and z-axes, respectively, and the rotational modes are called roll, pitch, and yaw for motions about these axes. In general, the motion of the structure will be a combination of movements in all of these directions and the velocity of a point on the surface, measured normal to the surface, may be written

$$V_n = \mathbf{U}.\mathbf{n} + \mathbf{\Omega}.(\mathbf{r} \times \mathbf{n}), \tag{1.31}$$

where $\mathbf{r}$ is the position vector of the point measured from the centre of rotation and $\mathbf{n}$ is a normal vector to the structure's surface directed out of the fluid. Here, the translational velocity vector

$$\mathbf{U} = (U_1, U_2, U_3) \tag{1.32}$$

has components corresponding to surge, sway, and heave motions, respectively, and the rotational velocity vector

$$\mathbf{\Omega} = (U_4, U_5, U_6) \tag{1.33}$$

has components corresponding to roll, pitch, and yaw, respectively. The normal velocity V_n given by (1.31) may be rewritten as

$$V_n = \sum_{\mu=1}^{6} U_\mu n_\mu, \tag{1.34}$$

where $\{n_\mu; \mu = 1, 2, 3\}$ are the x, y, z components of the unit normal to the structure defined by the direction cosines

$$n_1 = \cos(n, x), \quad n_2 = \cos(n, y), \quad n_3 = \cos(n, z), \tag{1.35}$$

while $\{n_\mu; \mu = 4, 5, 6\}$ are the corresponding components of $\mathbf{r} \times \mathbf{n}$. If a point on the structure's surface has coordinates (x, y, z) and (x', y', z') is the equilibrium position of the centre of rotation, then

$$\begin{aligned} n_4 &= (y - y')n_3 - (z - z')n_2, \\ n_5 &= (z - z')n_1 - (x - x')n_3, \\ n_6 &= (x - x')n_2 - (y - y')n_1. \end{aligned} \tag{1.36}$$

For time-harmonic motions, each velocity component may be written

$$U_\mu = \operatorname{Re}\left\{u_\mu \, \mathrm{e}^{-\mathrm{i}\omega t}\right\} \tag{1.37}$$

and hence by linear superposition the radiation velocity potential, defined in equation (1.27), may be decomposed as

$$\phi_{\mathrm{R}} = \sum_{\mu=1}^{6} u_\mu \phi_\mu \tag{1.38}$$

where u_μ is the complex amplitude of the oscillations in mode μ. The potential ϕ_μ describes the wave field due to oscillations in mode μ with unit velocity amplitude. The boundary condition (1.25) on the structural surface S_{B}, with V_n given by (1.34), is satisfied provided that

$$\frac{\partial \phi_\mu}{\partial n} = n_\mu \quad \text{on} \quad S_{\mathrm{B}}, \quad \mu = 1, 2, \ldots, 6. \tag{1.39}$$

Hence, ϕ_μ is a potential satisfying Laplace's equation within the fluid, the free-surface boundary condition (1.13), and the structural boundary condition (1.39). To obtain a unique solution, the radiated field must also satisfy a radiation condition in the form of (1.28) or (1.29) with $w = \phi_\mu$.

1.3.4   Hydrodynamic forces and moments

A structure immersed in water will be subject to forces due to the pressure from the surrounding fluid. From Bernoulli's equation (1.4), after linearization, the pressure at any point in the fluid, relative to that of the atmosphere, is

$$p = -\rho g z - \rho \frac{\partial \Phi}{\partial t}. \tag{1.40}$$

The first term is the pressure associated with the submergence of the measurement point, while the second term is the hydrodynamic pressure due to the motion of the fluid. Here, we are concerned only with the latter. The force on a structure is found by integrating the pressure multiplied by the unit normal vector over the surface of the structure and so, from (1.40), the μ component of the hydrodynamic force due to the fluid motion is given by

$$\mathcal{F}_\mu(t) = -\rho \iint\limits_{S_B} \frac{\partial \Phi}{\partial t} n_\mu \, \mathrm{d}S, \tag{1.41}$$

where the integral is taken over the mean wetted surface S_B for consistency with the linearization and n_μ is the generalized normal defined in §1.3.3. For time-harmonic motion of angular frequency ω the time-variation in the potential is removed as in (1.26) and (1.27) so that

$$\mathcal{F}_\mu(t) = \mathrm{Re}\left\{ F_\mu \, e^{-i\omega t} \right\}, \tag{1.42}$$

where from (1.41),

$$F_\mu = i\omega\rho \iint\limits_{S_B} \phi n_\mu \, \mathrm{d}S. \tag{1.43}$$

As explained previously, the velocity potential describing the fluid motion may be decomposed into two parts representing the scattered and radiated wave fields. The total hydrodynamic force on the structure due to these wave fields is decomposed similarly by writing

$$F_\mu = X_\mu + \sum_{\nu=1}^{6} u_\nu f_{\nu\mu}, \tag{1.44}$$

where the component of the so-called exciting force due to the scattered wave field is

$$X_\mu = i\omega\rho \iint\limits_{S_B} (\phi_I + \phi_D) n_\mu \, \mathrm{d}S \tag{1.45}$$

and the component of the force resulting from forced oscillations in mode ν is

$$f_{\nu\mu} = \mathrm{i}\omega\rho \iint\limits_{S_\mathrm{B}} \phi_\nu n_\mu \, \mathrm{d}S. \qquad (1.46)$$

It is conventional to further decompose the latter force into real and imaginary parts by writing

$$f_{\nu\mu} = \mathrm{i}\omega \left(a_{\nu\mu} + \frac{\mathrm{i}b_{\nu\mu}}{\omega} \right). \qquad (1.47)$$

The first term is that part of the force in phase with the acceleration in mode ν, and the second term that part in phase with the corresponding velocity. The real quantities $a_{\nu\mu}$ and $b_{\nu\mu}$ are termed the added mass and radiation damping (or just damping) coefficients and they may be considered as properties of the structure and, in general, are functions of frequency. From (1.46) and (1.47),

$$a_{\nu\mu} + \frac{\mathrm{i}b_{\nu\mu}}{\omega} = \rho \iint\limits_{S_\mathrm{B}} \phi_\nu n_\mu \, \mathrm{d}S. \qquad (1.48)$$

Further details of forces and moments and the equations of motion of a structure are given by Mei (1983, §7.2) and Newman (1977, Chapter 6).

1.3.5   Limitations of the inviscid, linear theory

The linearized theory assumes that the wavelength is much greater than the wave height (that is, the wave slope is small). The fluid motion generated by any moving structures must also be correspondingly small.

Figure 1.1 roughly indicates the region of validity of the inviscid linear theory described in this chapter. Here H is the wave height (trough to crest), D is a typical diameter of a structure and λ is the wavelength. When H/D is large, so that particle paths are long compared to the structural diameter, flow separation occurs and this changes substantially the flow from that predicted by the inviscid theory. This is the region marked 'viscous' in figure 1.1. When λ/D is large, so that the wavelength is much longer than typical structural dimensions, the wave field is little modified by the structure and wave diffraction is relatively unimportant. For moderate λ/D the wave field is significantly modified by the presence of the structure and wave diffraction effects must be

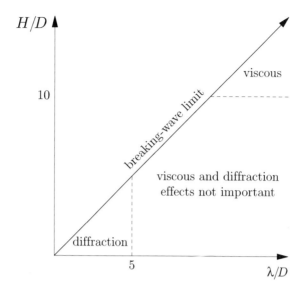

FIGURE 1.1
Regimes of importance for viscous and diffraction forces.
(Adapted from Figure 1.6 of O. M. Faltinsen, *Sea Loads on
Ships and Offshore Structures*, Cambridge University Press,
1990.)

included. As the fluid particle paths are now small compared to the
size of the structure, separation will not usually occur and the inviscid
theory may be used as viscous effects are confined to the thin boundary layers on the structure's surface. Thus, the primary application of
the theory is to the parameter regime marked 'diffraction' in Figure 1.1.
For further discussion of viscous effects see, for example, Sarpkaya and
Isaacson (1981, Chapter 6).

1.4   Reciprocity relations

There are some very general identities relating the quantities that
have been introduced which enhance our understanding of the physical
meaning of these quantities. They also provide checks that can be used

on analytical or numerical work and help reduce the effort required to calculate the quantities related by them. One way of deriving these identities is to use Green's theorem (see §4.2) which, for two harmonic potentials ϕ and ψ both satisfying the free-surface and bottom boundary conditions (1.13) and (1.14) (or equation 1.15), implies that

$$\iint_{S_B} \left(\phi \frac{\partial \psi}{\partial n} - \psi \frac{\partial \phi}{\partial n} \right) \mathrm{d}S + \lim_{X \to \infty} \iint_{S_X} \left(\phi \frac{\partial \psi}{\partial n} - \psi \frac{\partial \phi}{\partial n} \right) \mathrm{d}S = 0, \quad (1.49)$$

where S_B is the wetted surface of the structure, S_X is a vertical circular cylinder of radius X, and as usual the normal is directed out of the fluid region. In two dimensions S_X is replaced by two vertical lines at $x = \pm X$ and following the standard convention we write S_∞ for $\lim_{X \to \infty} S_X$.

By substitution of various radiation and scattering potentials in place of ϕ and ψ in (1.49) many useful results can be obtained. This was first done systematically by Newman (1976), though many of the results were known much earlier. The method of generating these so-called reciprocity relations is now standard and only brief details will be given here. A more thorough account, including references to the original derivations of the results, can be found in Mei (1983, §7.6), though some more recent results are described at the end of this section.

Suppose $\phi = \phi_\mu$, $\psi = \phi_\nu$, are two radiation potentials corresponding to two different modes of motion as defined in (1.38) and (1.39). It follows from (1.49) that the added mass and damping matrices given by (1.48) are symmetric. If we use $\psi = \overline{\phi_\nu}$, the complex conjugate of ϕ_ν, in (1.49) and make use of this symmetry and the radiation condition (1.29), we can relate the damping coefficient to the far-field form of the radiation potentials through

$$b_{\mu\nu} = \rho \omega k \iint_{S_\infty} \phi_\mu \overline{\phi_\nu} \, \mathrm{d}S. \quad (1.50)$$

If we restrict attention to the situation in which far from any structures the depth has a constant value, h, then in two dimensions we can assume (see equation 1.17) that

$$\phi_\mu \sim A_\mu^\pm \, \mathrm{e}^{\pm ikx} \frac{\cosh k(z + h)}{\cosh kh} \quad \text{as} \quad x \to \pm\infty, \quad (1.51)$$

for some constants $A_\mu^\pm$, which represent the amplitudes of waves radiated to $\pm\infty$, and then (1.50) gives

$$b_{\mu\nu} = \rho K c_g (A_\mu^- \overline{A_\nu^-} + A_\mu^+ \overline{A_\nu^+}), \quad (1.52)$$

where

$$c_g = \frac{d\omega}{dk} = \frac{\omega}{2k}\left(1 + \frac{2kh}{\sinh 2kh}\right) \tag{1.53}$$

is the group velocity for waves in water of depth h. (The group velocity is the velocity at which energy is propagated.)

In three dimensions we can assume that

$$\phi_\mu \sim A_\mu(\theta)\left(\frac{2}{\pi kr}\right)^{1/2} e^{ikr - i\pi/4} \frac{\cosh k(z+h)}{\cosh kh} \qquad \text{as} \quad r \to \infty \tag{1.54}$$

(this will be explained in §2.3.4 below), where $A_\mu(\theta)$ is some function which represents the angular dependence of the amplitude of the radiated waves in the far field and then the damping coefficients satisfy

$$b_{\mu\nu} = \frac{2\rho K c_g}{\pi k} \int_0^{2\pi} A_\mu(\theta)\overline{A_\nu(\theta)}\, d\theta. \tag{1.55}$$

It will be shown in §1.5 that this implies that the damping coefficients $b_{\mu\mu}$ are proportional to the energy radiated to infinity by a structure oscillating in mode μ.

Next we examine what happens if we use two scattering potentials in (1.49). Suppose $\phi = \phi^{(1)}$, $\psi = \phi^{(2)}$ with $\partial\phi^{(1)}/\partial n = \partial\phi^{(2)}/\partial n = 0$ on S_B. In two dimensions we can characterize such a potential by

$$\phi^{(j)} = \{A_j, B_j; C_j, D_j\}, \quad j = 1, 2, \tag{1.56}$$

implying that

$$\phi^{(j)} \sim -\frac{ig \cosh k(z+h)}{\omega \cosh kh} \times \begin{cases} A_j\, e^{ikx} + B_j\, e^{-ikx} & \text{as } x \to -\infty \\ C_j\, e^{ikx} + D_j\, e^{-ikx} & \text{as } x \to \infty. \end{cases} \tag{1.57}$$

Substitution of this into (1.49) gives the simple formula

$$A_1 B_2 - B_1 A_2 = C_1 D_2 - D_1 C_2. \tag{1.58}$$

A plane wave of unit amplitude incident from $x = -\infty$ and scattered by a structure is characterized by

$$\phi^{(1)} = \{1, R_1; T_1, 0\} \tag{1.59}$$

whilst if the incident wave is from $x = +\infty$ we have

$$\phi^{(2)} = \{0, T_2; R_2, 1\}. \tag{1.60}$$

where R_i and T_i, $i = 1, 2$, are known as reflection and transmission coefficients, respectively. Substituting these forms into (1.58) shows that

$$T_1 = T_2. \tag{1.61}$$

This is a remarkable result. It states that the transmission coefficient is independent of the direction of the incident wave regardless of the shape of the structure (or indeed the number of structures).

Instead of using ψ as above we can use $\psi = \overline{\phi^{(1)}} = \{\overline{R_1}, 1; 0, \overline{T_1}\}$ and then (1.58) gives

$$1 - |R_1|^2 = |T_1|^2 \tag{1.62}$$

which is a statement of the conservation of energy. Similarly $1 - |R_2|^2 = |T_2|^2$ and combining these equations with (1.61) shows that

$$|R_1| = |R_2|. \tag{1.63}$$

Thus the modulus of the reflection coefficient, though not necessarily the phase, is independent of the direction of the incident wave. (Equality of the phases occurs if the structure has symmetry about $x = 0$.) Information about the phase of the reflection and transmission coefficients can be obtained by taking $\phi = \phi^{(1)}$ and $\psi = \overline{\phi^{(2)}}$. If we write

$$T = |T| \, e^{i\delta}, \quad R_j = |R| \, e^{i\delta_j}, \quad j = 1, 2, \tag{1.64}$$

then (1.58) reduces to

$$\delta_1 + \delta_2 = 2\delta \pm \pi. \tag{1.65}$$

Similar results can be obtained for the three-dimensional case. A general scattering potential due to an incident wave of amplitude A, making an angle β_j with the positive x-axis, has the far-field form

$$\phi^{(j)} \sim -\frac{igA \cosh k(z + h)}{\omega \cosh kh} \left[e^{ikr \cos(\theta - \beta_j)} + \left(\frac{2}{\pi kr} \right)^{1/2} A^{(j)}(\theta) \, e^{ikr - i\pi/4} \right], \tag{1.66}$$

where we note that the normalized scattering amplitudes $A^{(j)}(\theta)$ (like R and T in the two-dimensional case) are non-dimensional, unlike the functions $A_\mu(\theta)$ defined in (1.54) (and $A_\mu^\pm$ defined in equation 1.51)

which have the dimensions of length. Two results that can be obtained
are

$$A^{(1)}(\beta_2 + \pi) = A^{(2)}(\beta_1 + \pi) \tag{1.67}$$

and

$$A^{(1)}(\beta_2) + \overline{A^{(2)}(\beta_1)} = -\frac{1}{\pi} \int_0^{2\pi} A^{(1)}(\theta) \overline{A^{(2)}(\theta)} \, d\theta. \tag{1.68}$$

Returning to the two-dimensional case, we can obtain further results
by considering the potential $\psi = -ig\omega^{-1}(\phi_\mu - \overline{\phi_\mu})$, where ϕ_μ is a radiation potential whose far field is given by (1.51). Since on the structure
$\partial\phi_\mu/\partial n = n_\mu$, which is real, it is clear that $\partial\psi/\partial n = 0$ on S_B and so ψ
is a scattering potential characterized by

$$\psi = \{-\overline{A_\mu^-}, A_\mu^-; A_\mu^+, -\overline{A_\mu^+}\}. \tag{1.69}$$

With $\phi = \phi^{(1)} = \{1, R_1; T, 0\}$, (1.58) gives

$$A_\mu^- + R_1\overline{A_\mu^-} + T\overline{A_\mu^+} = 0, \tag{1.70}$$

and with $\phi = \phi^{(2)} = \{0, T; R_2, 1\}$ we obtain

$$A_\mu^+ + R_2\overline{A_\mu^+} + T\overline{A_\mu^-} = 0. \tag{1.71}$$

These results are known as the Bessho-Newman relations. For the case
of a structure that is symmetric about a vertical plane making symmetric
(heave) oscillations we have

$$A_\mu^+ = A_\mu^- \equiv A_s = |A_s| \, e^{i\delta_s}, \tag{1.72}$$

say, whilst for such a structure making antisymmetric (surge or roll)
motions

$$A_\mu^+ = -A_\mu^- \equiv A_a = |A_a| \, e^{i\delta_a}, \tag{1.73}$$

say. Noting that $R_1 = R_2 \equiv R$ in this case, we see that the Bessho-Newman relations show that

$$R + T = -A_s/\overline{A_s} = -e^{2i\delta_s}, \qquad R - T = -A_a/\overline{A_a} = -e^{2i\delta_a}. \tag{1.74}$$

In three dimensions the Bessho-Newman relations are

$$\overline{A_\mu(\beta_j)} + A_\mu(\beta_j + \pi) + \frac{1}{\pi} \int_0^{2\pi} \overline{A_\mu(\theta)}A^{(j)}(\theta) \, d\theta = 0 \tag{1.75}$$

and they reduce to particularly simple expressions when the structure is axisymmetric (Davis 1976). For such a structure the scattered field due to an incident wave making an angle β_j with the positive x-axis is an even function of $\theta - \beta_j$ and so we can write the scattering amplitude as

$$A^{(j)}(\theta) = \sum_{m=0}^{\infty} \frac{\epsilon_m}{2} c_m \cos m(\theta - \beta_j) \qquad (1.76)$$

for some set of unknowns c_m, $m = 0, 1, 2, \ldots$. Here ϵ_m is the Neumann symbol, defined by $\epsilon_0 = 1$, $\epsilon_m = 2$, $m \geq 1$ and the factor $\epsilon_m/2$ is introduced for convenience. The angular dependence of the radiated waves in mode μ can be written in the form

$$A_\mu(\theta) = \sum_{m=0}^{\infty} \frac{\epsilon_m}{2} (-\mathrm{i})^m (a_m \cos m\theta + b_m \sin m\theta) \qquad (1.77)$$

and then (1.75) reduces to

$$a_m + (1 + c_m)\overline{a_m} = 0, \qquad m = 0, 1, 2, \ldots, \qquad (1.78)$$
$$b_m + (1 + c_m)\overline{b_m} = 0, \qquad m = 1, 2, 3, \ldots, \qquad (1.79)$$

from which it is clear that each of the coefficients c_m satisfies the equation $|1 + c_m| = 1$.

Finally, we can take $\phi = \phi_{\mathrm{I}} + \phi_{\mathrm{D}}$, the potential of the incident plus diffracted field on a fixed structure, and take $\psi = \phi_\mu$, the radiation potential for the forced motion of the same structure in mode μ. Since both ϕ_{D} and ϕ_μ describe outgoing waves at infinity, application of (1.49) results in

$$X_\mu = -\mathrm{i}\rho\omega \iint_{S_\infty} \left(\phi_{\mathrm{I}} \frac{\partial \phi_\mu}{\partial n} - \phi_\mu \frac{\partial \phi_{\mathrm{I}}}{\partial n} \right) \mathrm{d}S, \qquad (1.80)$$

which relates the exciting force in the μth direction, defined in (1.45), to the far field of the radiation potential for mode μ. In the two-dimensional case we assume (1.51) to hold and if the incident wave has amplitude A and is from $x = \pm\infty$ we obtain

$$X_\mu = -2\mathrm{i}\omega\rho c_g A A_\mu^{\pm}. \qquad (1.81)$$

In three dimensions, with the incident wave making an angle β with the positive x-axis we have

$$X_\mu(\beta) = -4\mathrm{i}\omega\rho k^{-1} c_g A A_\mu(\beta + \pi). \qquad (1.82)$$

These are known as the Haskind relations and they show that the exciting force in the μth direction is proportional to the amplitude of the waves radiated by the forced motion of the structure in mode μ, in the opposite direction to that of the incident wave.

All the reciprocity relations derived above relate far-field quantities, or quantities such as the damping coefficient which are defined in terms of integrals of near-field quantities. In two dimensions one can actually express the scattering potential at all points in the fluid in terms of the solutions to radiation problems. For example, for a structure which is symmetric about a vertical plane (and assuming that the scattering problem possesses a unique solution), McIver (1996b) showed that the scattering potential due to an incident wave of unit amplitude, $\phi^{(1)}$, defined in (1.59), can be expressed in terms of the solutions to two radiation problems. Thus if ϕ_{s} is the solution to the symmetric (heave) radiation problem and ϕ_{a} is the solution to the antisymmetric (surge or roll) radiation problem, with far-field amplitudes defined in (1.72) and (1.73), then

$$\frac{i\omega}{g}\phi^{(1)} = \frac{1}{2\overline{A}_{\text{a}}}(\phi_{\text{a}} - \overline{\phi_{\text{a}}}) - \frac{1}{2\overline{A}_{\text{s}}}(\phi_{\text{s}} - \overline{\phi_{\text{s}}}), \qquad (1.83)$$

provided that $A_{\text{a}} \neq 0$ and $A_{\text{s}} \neq 0$. Note that equation (1.74) follows immediately from the far-field form of this relation.

Reciprocity relations can also be derived for water-wave problems other than those described in this section. Thus Srokosz (1980) and Linton and Evans (1993b) derived relations applicable to a structure placed in a channel with vertical sides and Linton and McIver (1995) and Cadby and Linton (2000) derived relations (in two and three dimensions, respectively) applicable to a structure in a two-layer fluid consisting of a finite fluid layer of one density above an infinitely deep layer of greater density. Another situation in which reciprocity relations can be derived is described in §2.3.2. See also the bibliographical notes at the end of §6.3.

Finally, we note that reciprocity relations analogous to those described above for water-wave problems can also be derived for acoustic or electromagnetic waves, see for example Jones (1986, §1.35).

1.5 Energy of the fluid motion

Here the relationship between the added mass and damping coefficients, defined in (1.47), and the energy of the fluid motion in a forcing problem is considered. In particular, it is shown that the damping measures the energy radiated to the far field while the added mass is related to the energy of the near field motion.

Consider a radiation problem and let $\Phi(x, y, z, t)$ be the corresponding velocity potential. For simplicity a structure oscillating in a single mode of motion will be considered (the generalization to multiple modes of motion is given by Falnes and McIver 1985). Thus, as in (1.27) and (1.38),

$$\Phi(x, y, z, t) = \text{Re} \left\{ u\phi(x, y, z) \, e^{-i\omega t} \right\} \tag{1.84}$$

where u is the complex amplitude of the structures velocity. The corresponding added mass and damping coefficients, as defined in equation (1.47), will be denoted by $a_{\mu\mu}$ and $b_{\mu\mu}$, respectively.

1.5.1 Energy radiated to the far field

It has been shown in (1.50) that the damping coefficient is related to the far-field motion. In fact, it is proportional to the time-averaged energy flux of the waves radiating to the far field, as will now be shown.

The rate of working of the hydrodynamic force on an area element ΔS is $p_d \, \Delta S \, \partial\Phi/\partial n$, where $p_d = -\rho \, \partial\Phi/\partial t$ is the dynamic pressure and $\partial\Phi/\partial n$ is the outward normal component to ΔS of the fluid velocity. The total energy flux across S_∞ is therefore

$$E_f = -\rho \iint_{S_\infty} \frac{\partial\Phi}{\partial t} \frac{\partial\Phi}{\partial n} \, dS. \tag{1.85}$$

Now

$$\frac{\partial\Phi}{\partial t} \frac{\partial\Phi}{\partial n} = \text{Re} \left\{ -i\omega u\phi \, e^{-i\omega t} \right\} \text{Re} \left\{ u\frac{\partial\phi}{\partial n} e^{-i\omega t} \right\} \tag{1.86}$$

$$= \tfrac{1}{2}\omega \, \text{Im} \left\{ u^2\phi\frac{\partial\phi}{\partial n} e^{-2i\omega t} \right\} + \tfrac{1}{2}\omega \, \text{Im} \left\{ |u|^2\phi\frac{\partial\overline{\phi}}{\partial n} \right\} \tag{1.87}$$

where the result

$$\text{Re} \, z_1 \, \text{Re} \, z_2 = \tfrac{1}{2} \text{Re} \left\{ z_1 z_2 \right\} + \tfrac{1}{2} \text{Re} \left\{ z_1 \overline{z_2} \right\}, \tag{1.88}$$

which holds for any two complex numbers z_1 and z_2, has been used. The average energy flux over one period is therefore

$$\langle E_f \rangle = \frac{\omega}{2\pi} \int_0^{2\pi/\omega} E_f \, dt = -\tfrac{1}{2}\rho\omega|u|^2 \iint_{S_\infty} \mathrm{Im}\left\{\phi\frac{\partial\overline{\phi}}{\partial n}\right\} dS. \qquad (1.89)$$

By virtue of the radiation condition (1.29), as the radial coordinate $r \to \infty$

$$\phi\frac{\partial\overline{\phi}}{\partial r} \sim -\mathrm{i}k\phi\overline{\phi} \qquad (1.90)$$

and therefore from (1.50)

$$\langle E_f \rangle = \tfrac{1}{2}\rho\omega k|u|^2 \iint_{S_\infty} \phi\overline{\phi} \, dS = \tfrac{1}{2}|u|^2 b_{\mu\mu} \qquad (1.91)$$

and the time-averaged energy flux is proportional to the damping coefficient. In other words the damping coefficient measures the energy radiated away from the structure by the waves.

1.5.2   Potential and kinetic energy

The potential energy of a fluid column of horizontal cross-section ΔS due to the elevation η of the free surface is

$$\rho g \Delta S \int_0^\eta z \, dz, \qquad (1.92)$$

where the potential energy has been taken to be zero in the absence of waves. The total such potential energy of the fluid is therefore

$$V = \tfrac{1}{2}\rho g \iint_{S_F} \eta^2 dS. \qquad (1.93)$$

From (1.8),

$$\eta = \frac{\omega}{g} \, \mathrm{Re}\left\{\mathrm{i}u\phi(x,y,0)\,\mathrm{e}^{-\mathrm{i}\omega t}\right\} \qquad (1.94)$$

and hence by (1.88),

$$V = \frac{\rho\omega^2|u|^2}{4g} \iint_{S_F} \phi\overline{\phi} \, dS - \frac{\rho\omega^2}{4g} \, \mathrm{Re}\left\{u^2\,\mathrm{e}^{-2\mathrm{i}\omega t} \iint_{S_F} \phi^2 \, dS\right\} \qquad (1.95)$$

and so the time-averaged potential energy

$$\langle V \rangle = \frac{\rho \omega^2 |u|^2}{4g} \iint\limits_{S_F} \phi \overline{\phi} \, dS. \tag{1.96}$$

The total kinetic energy of the fluid motion over the fluid volume τ is

$$T = \tfrac{1}{2}\rho \iiint\limits_{\tau} \nabla \Phi \cdot \nabla \Phi \, d\tau \tag{1.97}$$

and hence from (1.88),

$$T = \tfrac{1}{4}\rho|u|^2 \iiint\limits_{\tau} \nabla \phi \cdot \nabla \overline{\phi} \, d\tau + \tfrac{1}{4}\rho \operatorname{Re} \left\{ u^2 \, e^{-2i\omega t} \iiint\limits_{\tau} \nabla \phi \cdot \nabla \phi \, d\tau \right\}. \tag{1.98}$$

As ϕ is a solution of Laplace's equation and by the divergence theorem

$$\iiint\limits_{\tau} \nabla \phi \cdot \nabla \overline{\phi} \, d\tau = \iiint\limits_{\tau} \nabla \cdot (\phi \nabla \overline{\phi}) \, d\tau = \iint\limits_{S} \phi \frac{\partial \overline{\phi}}{\partial n} \, dS, \tag{1.99}$$

where n is the outward normal coordinate to the surface S surrounding τ (the bed may be excluded as $\partial \overline{\phi}/\partial n$ is zero there). Thus, the time-averaged kinetic energy is

$$\langle T \rangle = \tfrac{1}{4}\rho|u|^2 \iint\limits_{S} \phi \frac{\partial \overline{\phi}}{\partial n} \, dS. \tag{1.100}$$

Now by (1.48), (1.96), and (1.50),

$$\iint\limits_{S} \phi \frac{\partial \overline{\phi}}{\partial n} \, dS = \iint\limits_{S_B} \phi \frac{\partial \overline{\phi}}{\partial n} \, dS + \frac{\omega^2}{g} \iint\limits_{S_F} \phi \overline{\phi} \, dS + \iint\limits_{S_\infty} \phi \frac{\partial \overline{\phi}}{\partial n} \, dS \tag{1.101}$$

$$= \frac{1}{\rho} \left(a_{\mu\mu} + \frac{i b_{\mu\mu}}{\omega} \right) + \frac{4}{\rho|u|^2} \langle V \rangle - \frac{i b_{\mu\mu}}{\rho \omega} \tag{1.102}$$

and so from (1.100),

$$\langle T \rangle - \langle V \rangle = \tfrac{1}{4} a_{\mu\mu} |u|^2. \tag{1.103}$$

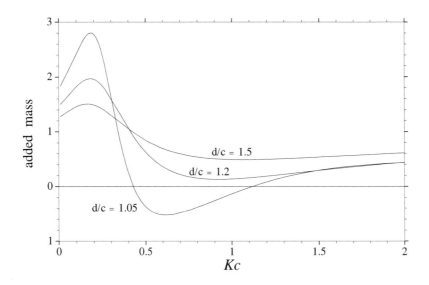

FIGURE 1.2
Non-dimensional added mass $a_{\mu\mu}/\rho\pi c^2$ of a horizontal cylinder
of radius c submerged to a depth d vs. frequency parameter Kc.

That is, the difference between the time-averaged kinetic and potential
energies is proportional to the added mass coefficient.

Consider the contribution to $\langle T \rangle$ from the region τ_R bounded by S_R,
a vertical circular at $r = R$, and S_∞. If R is sufficiently large for
the asymptotic form of the radiation potential (1.54) to apply then the
integrals over S_R and S_∞ cancel, leaving only the integral over the free
surface. It follows that in the region τ_R, $\langle T \rangle \approx \langle V \rangle$ and hence, from
(1.103), the added mass is related to the energy of the near-field motion.
When free-surface effects are dominant, so that V exceeds T, then the
added mass is negative. This is illustrated in Figure 1.2 for the two-
dimensional problem of an oscillating, submerged, horizontal circular
cylinder in deep water. The added masses in heave and sway are equal
for this geometry and are plotted here against a frequency parameter.
For the deeper submergences the added mass varies with frequency but
is always positive. If the submergence is reduced, so that the cylinder
interacts strongly with the free surface, a range of frequencies appears
for which the added mass is negative. This phenomenon is discussed in
detail by McIver and Evans (1984b).

Chapter 2

Eigenfunction expansions

The general theory in Chapter 1 shows that a linear water-wave/structure interaction problem requires the solution to Laplace's equation in the fluid domain subject to boundary conditions on the free surface, the bed, the structure or structures and, if the fluid is unbounded, at infinity. The techniques available for the solution of such a problem will depend on the geometry of the fluid domain and the geometries of the structures involved.

Laplace's equation separates in many different coordinate systems and if the geometry of the problem allows the boundary conditions to be separated also, progress can be made by utilizing this separation property. In this chapter we will investigate problems where the velocity potential can be written in terms of infinite series of separated eigenfunctions. This technique can thus be used when all the boundaries fit nicely into a particular coordinate system and so is fairly restricted in its application. Nevertheless it can be used to solve a number of basic problems simply and accurately.

Of course, realistic geometries will rarely have these nice properties but it will often be the case that away from the bodies under investigation the fluid domain will be such that eigenfunction expansions can be used. In such cases the region containing the body can be discretized using, say, a finite-element technique and this then matched to the analytic representation in the outer region. This 'hybrid-element' technique is described in detail in Mei (1983, §7.7).

2.1 Construction of vertical eigenfunctions

In the case where the depth h is finite and constant, equation (1.14) becomes

$$\frac{\partial \phi}{\partial z} = 0 \quad \text{on} \quad z = -h \tag{2.1}$$

and separable solutions of equation (1.12) may be sought in the form

$$\phi(x, y, z) = W(x, y)Z(z). \tag{2.2}$$

Substitution of this form for ϕ into Laplace's equation shows that

$$\frac{1}{W}\left(\frac{\partial^2 W}{\partial x^2} + \frac{\partial^2 W}{\partial y^2}\right) = -\frac{1}{Z}\frac{\mathrm{d}^2 Z}{\mathrm{d}z^2} = \alpha^2, \tag{2.3}$$

where the separation constant, or eigenvalue, α is to be determined by the boundary conditions. In this section attention will be paid to the form of Z, which will be referred to as a 'vertical eigenfunction'.

In view of (2.1) and (2.3), the general solution for $Z(z)$ may be conveniently written as

$$Z(z) = C \cos \alpha(z + h) + D \sin \alpha(z + h). \tag{2.4}$$

The bed condition (2.1) gives immediately that $D = 0$ and the free-surface condition (1.13) is satisfied provided α is a root of

$$K + \alpha \tan \alpha h = 0. \tag{2.5}$$

Equation (2.5) has an infinite sequence of positive real roots (see Figure 2.1) which will be denoted by $\{k_n; \; n = 1, 2, 3, \ldots\}$. There is also a sequence of negative roots $\{-k_n; \; n = 1, 2, 3, \ldots\}$, but these lead to exactly the same eigenfunctions and so need not be considered separately.

The nature of these roots can be important in applications. From Figure 2.1 it can be confirmed that $k_n h \in ((n - 1/2)\pi, n\pi)$ and it can be shown that as $n \to \infty$ for fixed Kh,

$$k_n h \sim n\pi - \frac{Kh}{n\pi} - \left(\frac{1}{Kh} - \frac{1}{3}\right)\left(\frac{Kh}{n\pi}\right)^3. \tag{2.6}$$

On the other hand, as $Kh \to \infty$ for fixed n we have the behaviour

$$k_n h \sim (n - \tfrac{1}{2})\pi \left(1 + \frac{1}{Kh} + \frac{1}{K^2 h^2}\right) \tag{2.7}$$

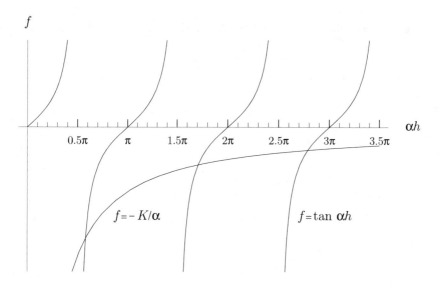

FIGURE 2.1
Intersections of $f = -K/\alpha$ and $f = \tan \alpha h$ showing the locations of the real roots αh of (2.5) in $\alpha h > 0$. These roots lie in the intervals $((n - 1/2)\pi, n\pi)$ for $n = 1, 2, 3, \ldots$.

and as $Kh \to 0$ for fixed n,

$$k_n h \sim n\pi - \frac{Kh}{n\pi} - \frac{(Kh)^2}{(n\pi)^3}. \tag{2.8}$$

In addition to the real roots of (2.5) there is also a pair of purely imaginary roots which will be denoted by $\alpha = \pm k_0$, where $k_0 = -ik$, say, and k is the positive root of the dispersion relation

$$K = k \tanh kh; \tag{2.9}$$

again it is sufficient to consider only the positive root. As explained later in §2.2, in regions which are unbounded in a horizontal direction this last root corresponds to propagating modes while each k_n $(n \geq 1)$ corresponds to a decaying or 'evanescent' mode.

Equations (2.5) and (2.9) are straightforward to solve numerically by standard methods. However, in applications, computational efficiency is often important and hence Newman (1990) and Chamberlain and Porter (1999) have obtained highly accurate approximations to the roots that

may then be refined by iteration. It is also worth noting that (2.5) and
(2.9) may be solved exactly in terms of integrals (Burniston and Siewert
1973), but this is computationally inefficient for practical calculations.

The vertical eigenfunctions are orthogonal because, for $m \neq n$,

$$
\int_{-h}^{0} \cos k_m(z+h) \cos k_n(z+h)\, dz
$$

$$
= \frac{k_m \sin k_m h \cos k_n h - k_n \sin k_n h \cos k_m h}{k_m^2 - k_n^2} = 0, \quad (2.10)
$$

where (2.5) has been used to simplify the result of the integration. It is
convenient to normalize the vertical eigenfunctions by writing

$$
\psi_n(z) = N_n^{-1} \cos k_n(z+h), \quad n = 0, 1, 2, \ldots \quad (2.11)
$$

and requiring

$$
\frac{1}{h} \int_{-h}^{0} [\psi_n(z)]^2\, dz = 1, \quad (2.12)
$$

so that

$$
N_n^2 = \frac{1}{2}\left(1 + \frac{\sin 2k_n h}{2 k_n h}\right). \quad (2.13)
$$

Two other forms for N_n that are sometimes useful are given by

$$
N_n^2 = \frac{1}{2}\left(1 - \frac{\sin^2 k_n h}{Kh}\right) = \frac{1}{2}\left(1 - \frac{K}{(K^2 + k_n^2)h}\right), \quad (2.14)
$$

which are equivalent to (2.13) by virtue of (2.5). With the above definitions, the orthogonality relations (2.10) and (2.12) may be written
as

$$
\frac{1}{h} \int_{-h}^{0} \psi_m(z)\psi_n(z)\, dz = \delta_{mn}, \quad (2.15)
$$

where δ_{mn} is the 'Kronecker delta' defined by $\delta_{mn} = 1$ if $m = n$, and
$\delta_{mn} = 0$ if $m \neq n$.

The problem of determining $Z(z)$ from (2.3) subject to boundary conditions on the free surface and bed is of standard Sturm-Liouville type
and so it follows (see e.g. Birkhoff and Rota 1989, Chapter 11, §3) that
the set $\{\psi_m;\ m = 0, 1, 2, \ldots\}$ is complete and any square integrable
function $f(z)$ defined on $(-h, 0)$ can be expanded as

$$
f(z) = \sum_{m=0}^{\infty} a_m \psi_m(z), \quad \text{where} \quad a_m = \frac{1}{h}\int_{-h}^{0} f(z)\psi_m(z)\, dz.
$$

2.2 Two-dimensional problems

In two-dimensional problems, where there is no dependence on y, equation (2.3) has the general solution

$$W(x, y) \equiv X(x) = A e^{-\alpha x} + B e^{\alpha x}, \qquad (2.16)$$

where A and B are constants to be determined from the boundary conditions. The general solution, or eigenfunction expansion, for the potential ϕ is obtained by superimposing all the possible modes to get

$$\phi(x, z) = \sum_{n=0}^{\infty} \left(A_n e^{-k_n x} + B_n e^{k_n x} \right) \psi_n(z). \qquad (2.17)$$

It follows from (1.10) that in this expression the term involving $\exp(\mathrm{i}kx)$ (recall that $k_0 = -\mathrm{i}k$) corresponds to a wave propagating towards large positive x while the term involving $\exp(-\mathrm{i}kx)$ corresponds to a wave propagating towards large negative x. Terms which decay as $x \to -\infty$ or as $x \to \infty$ are referred to as evanescent modes.

2.2.1 The wave-maker problem

Consider a semi-infinite tank of fluid occupying $x > 0$, $z \in (-h, 0)$. On $x = 0$ there is a wave-maker that oscillates with a specified velocity distribution at an angular frequency ω and hence generates waves that propagate towards large positive x. Denote the velocity potential for the flow by $\Phi(x, z, t)$ and suppose that the wave-maker boundary condition is

$$\frac{\partial \Phi}{\partial x} = U(z) \cos \omega t, \qquad (2.18)$$

where the real-valued function $U(z)$ is the imposed distribution of velocity on $x = 0$.

If the time dependence of Φ is factored out as in (1.10) the time-independent potential ϕ satisfies equations (1.12)–(1.14),

$$\frac{\partial \phi}{\partial x} = U(z) \quad \text{on} \quad x = 0 \qquad (2.19)$$

and a radiation condition

$$\frac{\partial \phi}{\partial x} - \mathrm{i}k\phi \to 0 \quad \text{as} \quad kx \to \infty \qquad (2.20)$$

that requires the solution to be bounded and any waves to be outgoing.

Application of the radiation condition (2.20) to the general solution (2.17) gives a form for the potential of

$$\phi(x, y) = \sum_{n=0}^{\infty} A_n \, e^{-k_n x} \, \psi_n(z). \tag{2.21}$$

From (2.19), it is required that

$$\left. \frac{\partial \phi}{\partial x} \right|_{x=0} = -\sum_{n=0}^{\infty} A_n k_n \psi_n(z) = U(z), \quad z \in (-h, 0), \tag{2.22}$$

and the unknown set of coefficients $\{A_n; \ n = 0, 1, 2, \dots\}$ can now be determined by exploiting the orthogonality properties (2.15) of the vertical eigenfunctions. In (2.22) multiply throughout by each of the set $\{\psi_m(z); \ m = 0, 1, 2, \dots\}$ in turn and integrate over the depth to get

$$A_m = -\frac{1}{k_m h} \int_{-h}^{0} U(z) \psi_m(z) \, dz. \tag{2.23}$$

The coefficient A_0 determines the amplitude and phase of the wave that propagates to large x.

For the special case of a vertical wave-maker in horizontal motion, so that $U(z) \equiv U$, a constant, the coefficients are

$$A_m = -\frac{U \sin k_m h}{k_m^2 h N_m}, \quad m = 0, 1, 2, \dots \tag{2.24}$$

and the amplitude and phase of the wave propagating away from the wave-maker can be obtained from

$$A_0 = -\frac{iU \sinh kh}{k^2 h N_0}. \tag{2.25}$$

The force on the wave-maker due to this surge motion can be determined in terms of the added mass and damping coefficients as described in §1.3.4, though the damping coefficient can be calculated more directly via the relation (1.52).

2.2.2   Forced oscillations of a rectangular tank

Another important two-dimensional problem is that of a partially-filled rectangular tank undergoing forced oscillations. Specifically we can consider a container bounded by $x = \pm b$, $z = -h$ containing liquid whose undisturbed free surface is at $z = 0$ and look at the case of forced time-harmonic horizontal oscillations of constant amplitude. When the time variation has been factored out in the manner of equation (1.10), the boundary conditions on the container walls become

$$\frac{\partial \phi}{\partial x} = 1 \quad \text{on} \quad x = \pm b \tag{2.26}$$

(c.f. equation 2.19 with $U(z) \equiv 1$) which implies that ϕ must be anti-symmetric in x.

One possible method of solution is to incorporate the antisymmetry in x into (2.17) so that

$$\phi(x, z) = \sum_{n=0}^{\infty} A_n \sinh k_n x \; \psi_n(z) \tag{2.27}$$

and then apply the boundary condition on $x = b$ to get

$$\sum_{n=0}^{\infty} A_n k_n \cosh k_n b \; \psi_n(z) = 1, \qquad z \in (-h, 0). \tag{2.28}$$

The orthogonality of the functions $\psi_n(z)$ can then be used to determine the coefficients A_n in the way described after equation (2.22) and we obtain

$$\phi = \sum_{n=0}^{\infty} \frac{\sin k_n h \sinh k_n x}{N_n k_n^2 h \cosh k_n b} \psi_n(z). \tag{2.29}$$

An alternative approach (see e.g. Graham and Rodriguez 1952) is to look for solutions of the form $\phi = x + \psi$ and then ψ satisfies the boundary-value problem

$$\nabla^2 \psi = 0 \qquad x \in (0, b), \; z \in (-h, 0), \tag{2.30}$$
$$K(x + \psi) = \psi_z \quad \text{on} \quad z = 0, \tag{2.31}$$
$$\psi_z = 0 \quad \text{on} \quad z = -h, \tag{2.32}$$
$$\psi = 0 \quad \text{on} \quad x = 0, \tag{2.33}$$
$$\psi_x = 0 \quad \text{on} \quad x = b, \tag{2.34}$$

where again the antisymmetry in x has been exploited. Equations (2.30) and (2.32)–(2.34) yield the expansion

$$\psi = \sum_{m=0}^{\infty} B_m \sin \mu_m x \cosh \mu_m(z + h), \tag{2.35}$$

where $\mu_m = (m + \frac{1}{2})\pi/b$ and then the free surface condition (2.31) gives

$$Kx = \sum_{m=0}^{\infty} B_m \sin \mu_m x \, (\mu_m \sinh \mu_m h - K \cosh \mu_m h). \qquad (2.36)$$

The orthogonality of the trigonometric functions can be used to determine the unknown coefficients B_m and we obtain

$$\phi = x + 2K \sum_{m=0}^{\infty} \frac{(-1)^m \sin \mu_m x \cosh \mu_m (z + h)}{\mu_m^2 b(K_m - K) \cosh \mu_m h}, \qquad (2.37)$$

where $K_m = \mu_m \tanh \mu_m h$. This form for the solution can be shown to be equivalent to (2.29) and both expressions reveal the resonances at $k = \mu_m$ (i.e. at $kb = (m + \frac{1}{2})\pi$, $m = 0, 1, 2, \dots$). The numbers K_m are the natural frequencies of antisymmetric oscillations in a rectangular container of width $2b$ and depth h.

2.3 Three-dimensional problems

Fully three-dimensional solutions of (2.3) will now be sought by the method of separation of variables. For rectangular geometries it is appropriate to write

$$W(x, y) = X(x)Y(y) \qquad (2.38)$$

and then we find that

$$\frac{1}{X}\frac{\mathrm{d}^2 X}{\mathrm{d}x^2} = k_n^2 - \frac{1}{Y}\frac{\mathrm{d}^2 Y}{\mathrm{d}y^2} = -\gamma^2, \qquad (2.39)$$

say. Hence

$$\frac{\mathrm{d}^2 X}{\mathrm{d}x^2} + \gamma^2 X = 0 \quad \text{and} \quad \frac{\mathrm{d}^2 Y}{\mathrm{d}y^2} - (k_n^2 + \gamma^2)Y = 0. \qquad (2.40)$$

2.3.1 Sloshing in a rectangular tank

Consider the problem of determining the natural frequencies of oscillation of fluid in a rectangular container (of length a and width b, say) with constant depth h. This is equivalent to looking for solutions to Laplace's equation in $x \in (0, a)$, $y \in (0, b)$, $z \in (-h, 0)$ which satisfy the bed boundary condition (1.14) together with wall boundary conditions

$$\frac{\partial \phi}{\partial x} = 0 \quad \text{on} \quad x = 0, a \quad \text{and} \quad \frac{\partial \phi}{\partial y} = 0 \quad \text{on} \quad y = 0, b. \tag{2.41}$$

The nature of these boundary conditions and the equations (2.40) shows that we must have $\gamma^2 > 0$ and $k_n^2 + \gamma^2 < 0$ which in turn shows that we must take $n = 0$ (since $k_0^2 = -k^2$) and then the solution is proportional to

$$\phi = \cos \frac{l\pi x}{a} \cos \frac{m\pi y}{b} \psi_0(z), \tag{2.42}$$

where l and m are arbitrary integers (not both zero) and

$$k^2 = \frac{l^2 \pi^2}{a^2} + \frac{m^2 \pi^2}{b^2}. \tag{2.43}$$

If one were to solve a forced motion problem for a rectangular tank, then resonances would be found at these natural frequencies just as for the two-dimensional case considered in §2.2.2 (see Graham and Rodriguez 1952).

2.3.2 Oblique waves

In certain situations it is necessary to consider geometries which do not vary in the y-direction, and to look for solutions which are periodic in y. In such cases we can look for solutions of (2.3) in the form

$$W(x, y) = X(x) e^{i \ell y} \tag{2.44}$$

from which we deduce (using equation 2.40 with $\gamma^2 = -k_n^2 - \ell^2$) that

$$X(x) = A e^{-\alpha_n x} + B e^{\alpha_n x}, \tag{2.45}$$

where $\alpha_n = (k_n^2 + \ell^2)^{1/2}$. Note that $\alpha_0 = (\ell^2 - k^2)^{1/2}$ and so (2.45) will represent a propagating wave if $\ell < k$. In this case we can write $\alpha_0 = -i(k^2 - \ell^2)^{1/2}$ and define an angle β by

$$(k^2 - \ell^2)^{1/2} = k \cos \beta, \qquad \ell = k \sin \beta. \tag{2.46}$$

A solution of (2.3) containing (2.45) with $n = 0$ would then be of the form

$$\phi(x, y, z) = \varphi(x, z)\, e^{i\ell y}, \tag{2.47}$$

where

$$\varphi = (A\, e^{ikx\cos\beta} + B\, e^{-ikx\cos\beta})\psi_0(z). \tag{2.48}$$

The term involving $\exp(ikx\cos\beta)$ represents a wave making an angle β with the positive x-axis while the term involving $\exp(-ikx\cos\beta)$ represents a wave making an angle $\pi - \beta$ with the positive x-axis. The two-dimensional case can be recovered by setting $\beta = 0$.

Since ϕ satisfies Laplace's equation throughout the fluid domain, the reduced potential φ satisfies the two-dimensional modified Helmholtz equation

$$\nabla^2\varphi - \ell^2\varphi = 0 \quad \text{in the fluid}, \tag{2.49}$$

and has the eigenfunction expansion

$$\varphi(x, z) = \sum_{n=0}^{\infty} \left(A_n\, e^{-\alpha_n x} + B_n\, e^{\alpha_n x}\right)\psi_n(z). \tag{2.50}$$

Although the fluid motion is three-dimensional, the boundary-value problem that we need to solve is actually two-dimensional. Reciprocity relations, similar to those derived in §1.4 for the strictly two-dimensional case, can be derived in this case also. The appropriate radiation problems no longer correspond to rigid body motions, however, but to forced motion problems where the forcing has an $\exp(i\ell y)$ dependence. See Green (1971), Bolton and Ursell (1973), and Garrison (1985) for more details.

2.3.3   Scattering by a symmetric obstacle

Suppose a wave making an angle β (which might be zero) with the positive x-axis is scattered by an obstacle which is in the form of an infinite cylinder whose generators are parallel to the y-axis and whose cross-section is symmetric about $x = 0$. We can look for a solution of the form $\varphi(x, z)\exp(i\ell y)$, where $\varphi(x, z)$ satisfies (2.49) together with the free-surface and bed boundary conditions, (1.13) and (1.14), and we must have $\partial\varphi/\partial n = 0$ on the body boundary. The appropriate radiation condition for this problem shows that

$$\varphi \sim \begin{cases} (\mathrm{e}^{\mathrm{i}kx\cos\beta} + R\,\mathrm{e}^{-\mathrm{i}kx\cos\beta})\psi_0(z) & \text{as} \quad x \to -\infty \\ T\,\mathrm{e}^{\mathrm{i}kx\cos\beta}\,\psi_0(z) & \text{as} \quad x \to \infty, \end{cases} \tag{2.51}$$

where R and T are the unknown reflection and transmission coefficients, respectively.

Another way of formulating this problem is to split φ into a part φ^+ which is symmetric about $x = 0$ and an antisymmetric part φ^- such that $\varphi = \varphi^+ + \varphi^-$. If we denote that part of the line $x = 0$, $z \in (-h, 0)$ which does not contain the obstacle by L, then

$$\frac{\partial\varphi^+}{\partial x} = 0 \quad \text{on} \quad L, \qquad \varphi^- = 0 \quad \text{on} \quad L. \tag{2.52}$$

We then only need consider the region $x < 0$ since the solutions can be extended into the whole fluid region using the symmetry relations $\varphi^+(x, z) = \varphi^+(-x, z)$ and $\varphi^-(x, z) = -\varphi^-(-x, z)$.

The appropriate behaviour as $x \to -\infty$ for these new potentials is

$$\varphi^\pm \sim \frac{1}{2}\left(\mathrm{e}^{\mathrm{i}kx\cos\beta} + R^\pm\,\mathrm{e}^{-\mathrm{i}kx\cos\beta}\right)\psi_0(z), \tag{2.53}$$

and if we can solve for R^+ and R^-, the reflection and transmission coefficients for the full problem can be recovered from the equations

$$R = \frac{1}{2}\left(R^+ + R^-\right), \qquad T = \frac{1}{2}\left(R^+ - R^-\right). \tag{2.54}$$

The reduction of a problem posed over the whole range $x \in (-\infty, \infty)$ to two problems each posed on $x < 0$ often leads to a considerable saving of effort.

2.3.4   Cylindrical polar coordinates

Here the general eigenfunction expansion for three-dimensional problems will be constructed in terms of cylindrical polar coordinates (r, θ, z) where the polar coordinates (r, θ) are defined as in (1.16) through

$$x = r\cos\theta \quad \text{and} \quad y = r\sin\theta. \tag{2.55}$$

In terms of these coordinates, equation (2.3) for $\mathcal{W}(r, \theta) \equiv W(x, y)$ is

$$\frac{1}{r}\frac{\partial}{\partial r}\left(r\frac{\partial\mathcal{W}}{\partial r}\right) + \frac{1}{r^2}\frac{\partial^2\mathcal{W}}{\partial\theta^2} = k_n^2\mathcal{W}. \tag{2.56}$$

Look for separable solutions in the form

$$\mathcal{W}(r,\theta) = R(r)\Theta(\theta) \tag{2.57}$$

so that from equation (2.56),

$$\frac{r^2}{R}\left[\frac{1}{r}\frac{d}{dr}\left(r\frac{dR}{dr}\right) - k_n^2 R\right] = -\frac{1}{\Theta}\frac{d^2\Theta}{d\theta^2} = \beta^2, \tag{2.58}$$

say. The general solution for Θ is

$$\Theta(\theta) = A\cos\beta\theta + B\sin\beta\theta, \tag{2.59}$$

for some constants A and B. In many applications, for example those involving a fluid domain exterior to a finite structure, the requirement that the potential be continuous implies that β must be an integer, m, say. The equation governing $R(r)$ is then

$$r\frac{d}{dr}\left(r\frac{dR}{dr}\right) - (m^2 + k_n^2 r^2)R = 0, \quad m = 0, 1, 2, \ldots, \tag{2.60}$$

which is the modified Bessel's equation (A.3). Thus

$$R(r) = CI_m(k_n r) + DK_m(k_n r), \quad m, n = 0, 1, 2, \ldots, \tag{2.61}$$

for some constants C and D, where I_m and K_m denote the modified Bessel functions of the first and second kind, respectively, and of order m. The general solution for ϕ is

$$\phi = \sum_{m=0}^{\infty}\sum_{n=0}^{\infty}[A_{mn}\cos m\theta + B_{mn}\sin m\theta]$$

$$\times [C_{mn}I_m(k_n r) + D_{mn}K_m(k_n r)]\,\psi_n(z). \tag{2.62}$$

Since $k_0 = -ik$, it follows from (A.4) and (A.5) that

$$I_m(k_0 r) = (-i)^m J_m(kr) \quad \text{and} \quad K_m(k_0 r) = \tfrac{1}{2}\pi i^{m+1}H_m^{(1)}(kr), \tag{2.63}$$

where J_m is a Bessel function of the first kind and $H_m^{(1)}$ is a Hankel function of the first kind, both of order m. The behaviour of $J_m(kr)$ and $H_m^{(1)}(kr)$ for large argument, given by (A.10) and (A.12), shows that these terms correspond to propagating modes and it follows from (A.2) that the combination of radial functions

$$C_{m0}I_m(k_0 r) + D_{m0}K_m(k_0 r) \tag{2.64}$$

has two useful alternative representations; these are

$$C'_{m0} J_m(kr) + D'_{m0} H_m^{(1)}(kr) \tag{2.65}$$

and

$$C''_{m0} J_m(kr) + D''_{m0} Y_m(kr). \tag{2.66}$$

Here Y_m denotes the Bessel function of the second kind and of order m. From (A.12) and the radiation condition (1.29), a solution which represents outgoing waves will be of the form

$$\phi = \sum_{m=0}^{\infty} (A_m \cos m\theta + B_m \sin m\theta) H_m^{(1)}(kr)\psi_0(z) \tag{2.67}$$

$$\sim f(\theta) \left(\frac{2}{\pi kr}\right)^{1/2} e^{ikr - i\pi/4} \psi_0(z) \quad \text{as} \quad kr \to \infty, \tag{2.68}$$

for some function f. Thus, the form (2.65) is suited to representations of a potential for scattering or radiation problems where a radiation condition is imposed. The form (2.66) is suited to representations of a potential in regions of finite horizontal extent.

The other terms in (2.62) either become unbounded or decay as the radial coordinate increases. From (A.8) it follows that the terms $I_m(k_n r)$, $n > 0$, all increase exponentially as $k_n r \to \infty$ and so will not be present in the expansion of a potential in a region which extends to infinity. On the other hand (A.9) shows that the terms $K_m(k_n r)$, $n > 0$, all decrease exponentially as $k_n r \to \infty$. These latter terms are the evanescent modes.

2.3.5 Sloshing in a cylindrical tank

Consider the problem of determining the natural frequencies of oscillation of fluid in a cylindrical container with a vertical axis, of radius a and height h. These will be the frequencies at which resonances occur if the container is forced to undergo oscillatory motions (for a solution to the problem of forced horizontal motions, see Isaacson and Subbiah 1991). In our case we wish to solve $\nabla^2 \phi = 0$ within the container together with the free-surface condition (1.13), the bed boundary condition (2.1), and

$$\frac{\partial \phi}{\partial r} = 0 \quad \text{on} \quad r = a. \tag{2.69}$$

Any solution can be expressed in the general form (2.62) and the orthogonality of the functions $\cos m\theta$ and $\sin m\theta$ on $(0, 2\pi)$ and $\psi_n(z)$ on $(-h, 0)$ implies that the component of $\partial\phi/\partial r$ for each n and m must vanish separately. Since the solution must be regular at $r = 0$, (A.14) and (A.15) show that all the coefficients of $K_m(k_n r)$ must be zero and since $I'_m(x) > 0$ for all $x > 0$ (Abramowitz and Stegun 1965, §9.6), the only possibility is

$$\phi(r, \theta, z) = (A_m \cos m\theta + B_m \sin m\theta) J_m(kr) \psi_0(z) \qquad (2.70)$$

for any integer m. Application of (2.69) shows that $J'_m(ka) = 0$ and hence that $ka = j'_{mn}$, where j'_{mn} is the nth zero of J'_m. Properties of such zeros are given in (A.23)–(A.27).

This example can easily be extended to cover the case of an annular cylinder with fluid occupying the region $r \in (b, a)$, $z \in (-h, 0)$. Since we no longer require regularity at the origin, the r dependence of the solution has the more general form (2.66) and the condition of zero normal flow on $r = a, b$ shows that the eigenvalues ka are given by the solutions for x of

$$J'_m(x) Y'_m(xb/a) - J'_m(xb/a) Y'_m(x) = 0 \qquad (2.71)$$

(see, for example, McLachlan 1954, §2.64).

2.4   The Helmholtz equation

In water-wave scattering problems where the depth is constant, equation (1.17) shows that an incident plane wave takes the form (up to an arbitrary multiplicative constant)

$$\phi_{\text{I}} = e^{ikr \cos(\theta - \beta)} \psi_0(z) \qquad (2.72)$$

for some angle β. If the scatterer has constant cross-section and extends throughout the water depth, the orthogonality of the functions $\{\psi_n(z); \ n = 0, 1, 2, \dots\}$ implies that the diffracted wave will also be proportional to $\psi_0(z)$. The potential for the problem may thus be decomposed as

$$\phi(r, \theta, z) = \varphi(r, \theta) \psi_0(z) \qquad (2.73)$$

and, since ϕ satisfies Laplace's equation (1.12), φ satisfies the Helmholtz equation

$$\nabla^2 \varphi + k^2 \varphi = 0 \tag{2.74}$$

throughout the fluid domain, and ∇^2 now denotes the Laplacian operator in terms of the horizontal coordinates only.

Such problems have direct analogues in the theory of acoustics and electromagnetic waves. Thus in linear acoustics, the pressure fluctuation p satisfies the wave equation

$$\nabla^2 p = \frac{1}{c^2} \frac{\partial^2 p}{\partial t^2}, \tag{2.75}$$

where c is the speed of sound and if one assumes a time harmonic variation of the form $p = \mathrm{Re}\{\varphi \exp(-i\omega t)\}$, one obtains (2.74) where now $k = \omega/c$. The same type of reduction occurs in Maxwell's theory of electromagnetic waves, where the electric potential for a disturbance in free space satisfies the wave equation (2.75) but with c now the speed of light; see Jones (1986) for more details.

Many of the examples given in this and subsequent chapters involve the solution of the Helmholtz equation and the above remarks demonstrate that techniques developed for such problems have applications in a number of physical contexts.

2.4.1 Scattering by a vertical circular cylinder

Consider the scattering of an incident plane wave of wavenumber k by a bottom-mounted vertical circular cylinder standing in water of constant depth h. Without loss of generality we can assume that the axis of the cylinder coincides with the z-axis and that the incident wave is from $x = -\infty$. The potential of the incident wave is written as

$$\phi_{\mathrm{I}} = e^{ikx} \, \psi_0(z) \tag{2.76}$$

and the total potential for the problem may be decomposed according to $\phi(r, \theta, z) = \varphi(r, \theta)\psi_0(z)$, where φ satisfies the Helmholtz equation (2.74) throughout the fluid domain.

The potential can be further decomposed as $\varphi = \varphi_{\mathrm{I}} + \varphi_{\mathrm{D}}$ and in cylindrical polar coordinates we have

$$\varphi_{\mathrm{I}} = e^{ikr\cos\theta} = \sum_{m=0}^{\infty} \epsilon_m i^m J_m(kr) \cos m\theta \tag{2.77}$$

(Abramowitz and Stegun 1965, 9.1.42–43), where ϵ_m is the Neumann symbol defined on page 17. Since the diffracted wave φ_D must behave like an outgoing wave as $kr \to \infty$ the result (2.67) means that we can write

$$\varphi_D = \sum_{m=0}^{\infty} a_m \epsilon_m i^m H_m^{(1)}(kr) \cos m\theta, \tag{2.78}$$

where the complex coefficients a_m are to be determined and the orthogonality of the functions in θ has been used to eliminate terms in $\sin m\theta$. Application of the body boundary condition

$$\frac{\partial \varphi_D}{\partial r} = -\frac{\partial \varphi_I}{\partial r} \quad \text{on} \quad r = a \tag{2.79}$$

and the orthogonality of the functions $\cos m\theta$ on $(0, 2\pi)$ gives

$$a_m = -\frac{J_m'(ka)}{H_m^{(1)'}(ka)}. \tag{2.80}$$

Thus

$$\varphi(r, \theta) = \sum_{m=0}^{\infty} \epsilon_m i^m \left(J_m(kr) - \frac{J_m'(ka)}{H_m^{(1)'}(ka)} H_m^{(1)}(kr) \right) \cos m\theta. \tag{2.81}$$

In practical applications a quantity of interest is the value of ϕ (which is proportional to the pressure) on the cylinder and this has a particularly simple form due to the Wronskian relations satisfied by the Bessel functions, equation (A.7). We find that the total potential on the cylinder surface is given by

$$\phi(a, \theta, z) = \frac{2\psi_0(z)}{\pi ka} \sum_{m=0}^{\infty} \frac{\epsilon_m i^{m+1}}{H_m^{(1)'}(ka)} \cos m\theta. \tag{2.82}$$

Mei (1983, §7.5) gives more information on how this can be used to calculate quantities such as the forces on the cylinder.

Bibliographical notes

The specific example of water-wave scattering by a vertical circular cylinder in finite depth seems to have been examined first by Omer and Hall (1949), though the work of MacCamy and Fuchs (1954) is perhaps better known. The radiation problem for the forced horizontal oscillations of a vertical cylinder, for which a reduction to the two-dimensional Helmholtz equation is no longer possible, was solved in terms of an eigenfunction expansion by Isaacson, Mathai, and Mihelcic (1990) and comparisons were made with experimental data. A different form for the solution of the radiation problem involving an integral representation was given by McIver (1994b).

In the context of acoustics the history of the problem described in this section goes back to the 19th century with the work of Lord Rayleigh in 1878 on the interaction of sound waves with cylinders whose diameter is small compared to wavelength (Rayleigh 1945, §343). Indeed, the problem of the interaction of a sound wave with an arbitrary finite array of circular cylinders was solved in 1913, as will be described in §6.1 below.

2.5 Matched eigenfunction expansions

There are very few geometries for which the solution can be written in the form of a single eigenfunction expansion as in the previous section. One technique that may be applicable for a more general geometry is to represent the solution as a different expansion in different regions and then to match these expansions across the boundaries between these regions.

2.5.1 Scattering by a vertical barrier

Consider the diffraction of an oblique incident plane wave making an angle β with the positive x-axis by a thin vertical barrier which occupies $x = 0$, $z \in [-d, 0]$, $y \in (-\infty, \infty)$ in water of uniform depth h. The theory presented in §2.3.2 shows that the total velocity potential for the scattering problem can be written $\varphi(x, z) \exp(i\ell y)$, where $\ell = k \sin \beta$ and then $\varphi(x, z)$ satisfies the modified Helmholtz equation (2.49), the bed and free-surface boundary conditions (1.13) and (2.1), together with

$$\frac{\partial \varphi}{\partial x} = 0 \quad \text{on} \quad x = 0, \ z \in (-d, 0). \tag{2.83}$$

The appropriate radiation conditions are

$$\varphi \sim \begin{cases} \left(\mathrm{e}^{\mathrm{i}\alpha x} + R\,\mathrm{e}^{-\mathrm{i}\alpha x}\right)\psi_0(z) & \text{as} \quad x \to -\infty \\ T\,\mathrm{e}^{\mathrm{i}\alpha x}\,\psi_0(z) & \text{as} \quad x \to \infty, \end{cases} \tag{2.84}$$

where $\alpha = k\cos\beta$ and R and T are the reflection and transmission coefficients which we wish to determine.

In order to complete the description of the problem we must specify the behaviour of the solution in the vicinity of the bottom of the barrier. If we consider a small fluid region near the barrier tip, then the potential φ is essentially governed by Laplace's equation because of the rapid variations in this region. If we introduce polar coordinates (r,θ) with origin at the barrier tip $(x, z) = (0, -d)$, so that the sides of the barrier correspond to $\theta = 0$ and $\theta = 2\pi$, then the boundary condition on the barrier walls is $\partial\varphi/\partial\theta = 0$ on $\theta = 0, 2\pi$. The solution must therefore satisfy $\varphi \sim c + r^{1/2}\cos(\theta/2)$ as $r \to 0$, where c is a constant. In other words the singularity in the velocity field for the full problem must be governed by the relation

$$\nabla\varphi = O(r^{-1/2}) \quad \text{as} \quad r \to 0. \tag{2.85}$$

From (2.50) and (2.84), in $x < 0$ we can expand the potential as the eigenfunction series

$$\varphi = \left(\mathrm{e}^{\mathrm{i}\alpha x} + R\,\mathrm{e}^{-\mathrm{i}\alpha x}\right)\psi_0(z) + \sum_{n=1}^{\infty} A_n\,\mathrm{e}^{\alpha_n x}\,\psi_n(z), \tag{2.86}$$

where $\alpha_n = (k_n^2 + \ell^2)^{1/2}$. Similarly, in $x > 0$,

$$\varphi = T\,\mathrm{e}^{\mathrm{i}\alpha x}\,\psi_0(z) + \sum_{n=1}^{\infty} B_n\,\mathrm{e}^{-\alpha_n x}\,\psi_n(z). \tag{2.87}$$

The complex numbers R, T, A_n, B_n, $n \geq 1$, must be determined by matching these representations for φ across the plane $x = 0$.

Physically it makes sense to match the fluid pressure and velocity across $x = 0$, $z \in (-h, -d)$, which corresponds to imposing continuity of φ and φ_x across this region. Mathematically, these conditions imply that the potential $\varphi(x, z)\exp(\mathrm{i}\ell y)$ in $x < 0$ is the harmonic continuation of that in $x > 0$ (see, for example, Kellogg 1953, Chapter X, §5, Theorem VI). Bearing in mind the boundary condition (2.83), it is clear that

φ_x must be continuous across the whole interval $x = 0$, $z \in (-h, 0)$ and so

$$i\alpha(1 - R)\psi_0(z) + \sum_{n=1}^{\infty} \alpha_n A_n \psi_n(z)$$

$$= i\alpha T \psi_0(z) - \sum_{n=1}^{\infty} \alpha_n B_n \psi_n(z), \qquad z \in (-h, 0). \quad (2.88)$$

The orthogonality of the depth eigenfunctions then shows that

$$1 - R = T \quad \text{and} \quad A_n = -B_n, \quad n \geq 1. \quad (2.89)$$

The other available information can then be expressed by the equations

$$\psi_0(z) + \sum_{n=0}^{\infty} A_n \psi_n(z) = 0 \quad z \in (-h, -d), \quad (2.90)$$

$$\sum_{n=0}^{\infty} \alpha_n A_n \psi_n(z) = 0 \quad z \in (-d, 0), \quad (2.91)$$

where we have written $A_0 = R - 1$ and $\alpha_0 = -i\alpha$. The first of these equations follows from the continuity of φ across $x = 0$, $z \in (-h, -d)$ after using (2.89) and the second follows from (2.83).

Another way of deriving these equations is to split the problem into a symmetric and an antisymmetric part as described in §2.3.3. The symmetric potential φ^+ satisfies $\partial \varphi^+ / \partial x = 0$ on $x = 0$ and the solution is simply $\varphi^+ = \cos \alpha x \, \psi_0(z)$. The antisymmetric part φ^- satisfies the condition $\varphi^- = 0$ on $x = 0$, $z \in (-h, -d)$ together with (2.83) and we only need consider the region $x < 0$, in which φ^- can be expanded as

$$\varphi^- = \frac{1}{2} \left(e^{i\alpha x} + R^- e^{-i\alpha x} \right) \psi_0(z) + \sum_{n=1}^{\infty} A_n e^{\alpha_n x} \psi_n(z) \quad (2.92)$$

so that $\varphi = \varphi^+ + \varphi^-$ and from (2.86),

$$R = \frac{1}{2} \left(1 + R^- \right). \quad (2.93)$$

The boundary condition $\varphi^- = 0$ on $x = 0$, $z \in (-h, -d)$ then reduces to (2.90) and $\partial \varphi^- / \partial x = 0$ on $x = 0$, $z \in (-d, 0)$ implies (2.91).

Equations (2.90)–(2.91) can be converted into an infinite system of equations for A_n, $n \geq 0$ in a variety of different ways. One possibility is to multiply both equations by the set of functions $\psi_m(z)$, $m \geq 0$, integrate over the appropriate range and then add the resulting equations. This leads to

$$\alpha_m h A_m + \sum_{n=0}^{\infty} [c_{nm}(1 - \alpha_n h)] A_n = -c_{0m}, \quad m = 0, 1, 2, \ldots, \quad (2.94)$$

where

$$c_{nm} = \frac{1}{h} \int_{-h}^{-d} \psi_n(z)\psi_m(z)\,\mathrm{d}z = \delta_{nm} - \frac{1}{h} \int_{-d}^{0} \psi_n(z)\psi_m(z)\,\mathrm{d}z, \quad (2.95)$$

see (2.15).

The system (2.94) can be solved by truncating and numerically solving an $N \times N$ system of equations. However, because of the singularity in the velocity near the tip of the plate revealed by (2.85), the convergence as N increases is very slow. McIver (1985b), who considered the scattering of waves by two barriers by the above method, suggested that it was necessary to take N as large as 400 and then use linear extrapolation to achieve satisfactory results. In order to derive a system of equations with much better convergence characteristics it is necessary to use a solution procedure which accurately models the singular behaviour near the end of the plate and this can be done using the Galerkin method also used in §4.7.4 below.

First we introduce the function

$$U(z) = \left.\frac{\partial \varphi}{\partial x}\right|_{x=0} = \sum_{n=0}^{\infty} \alpha_n A_n \psi_n(z), \quad (2.96)$$

representing the horizontal fluid velocity on $x = 0$. This function can be expanded in terms of the orthogonal eigenfunctions $\psi_n(z)$ as

$$U(z) = \sum_{n=0}^{\infty} U_n \psi_n(z), \quad (2.97)$$

where, since $U(z) = 0$ for $z \in (-d, 0)$,

$$U_n = \alpha_n A_n = \frac{1}{h} \int_{-h}^{-d} U(z)\psi_n(z)\,\mathrm{d}z, \quad n = 0, 1, 2, \ldots. \quad (2.98)$$

It is convenient to introduce a function $u(z)$ defined by $U(z) = -Ru(z)$ and then the equation representing continuity of pressure across the gap under the barrier, (2.90), can be written, using (2.98) for $n \geq 1$, as

$$Ku \equiv \int_{-h}^{-d} u(t)K(z,t)\,\mathrm{d}t = \psi_0(z), \qquad (2.99)$$

where

$$K(z,t) = \sum_{n=1}^{\infty} \frac{1}{\alpha_n h}\psi_n(t)\psi_n(z). \qquad (2.100)$$

The integral operator K defined by (2.99) and (2.100) is real, linear, symmetric, and positive definite and it follows that the solution $u(z)$ to the integral equation (2.99) is real. Once (2.99) has been solved for u, then the relation

$$\int_{-h}^{-d} u(z)\psi_0(z)\,\mathrm{d}z = \mathrm{i}\alpha h(R-1)/R, \qquad (2.101)$$

which follows from (2.98) with $n = 0$, may be used to determine the reflection coefficient R. Many boundary-value problems may be formulated in terms of an integral equation of the form (2.99) and with a condition corresponding to (2.101). The advantages of such a formulation are described in §8.1.

To obtain a numerical solution to (2.99) efficiently, we approximate $u(z)$ for $z \in (-h, -d)$ by writing it as

$$u(z) = \sum_{n=0}^{N} a_n u_n(z), \qquad (2.102)$$

where the functions $u_n(z)$ are chosen to correctly model the singular behaviour of the fluid velocity near to the submerged end of the barrier. Since, from (2.85), $u(z) = O(|z+d|^{-1/2})$ as $z \to -d$ from below and, from (2.1), $u'(z) = 0$ on $z = -h$, we choose to write

$$u_n(z) = \left((h-d)^2 - (h+z)^2\right)^{-1/2} T_{2n}\left(\frac{h+z}{h-d}\right), \qquad (2.103)$$

where $T_n(\cos\theta) = \cos n\theta$ is a Chebyshev polynomial. This expression has the required form and the choice of Chebyshev polynomials allows certain integrals that occur below to be evaluated explicitly. The representation (2.102) is then substituted into (2.99) and (2.101), the resulting equations are multiplied by each of the set of functions $u_m(z)$ and integrated over $(-h, -d)$. This yields

$$\sum_{n=0}^{N} a_n K_{mn} = F_{m0}, \quad m = 0, 1, 2, \ldots \qquad (2.104)$$

and

$$\sum_{n=0}^{N} a_n F_{n0} = i\alpha h(R-1)/R, \qquad (2.105)$$

where

$$K_{mn} = \sum_{s=1}^{\infty} \frac{1}{\alpha_s h} F_{ms} F_{ns} \qquad (2.106)$$

and

$$F_{mn} = \int_{-h}^{-d} u_m(z) \psi_n(z) \, dz = \frac{\pi}{2} (-1)^m N_n^{-1} J_{2m}\left(k_n(h-d)\right). \quad (2.107)$$

Here J_{2m} is a Bessel function of the first kind and the last integral has been evaluated using Gradshteyn and Ryzhik (1980, eqn 7.355(2)). The infinite system of equations (2.104) has excellent convergence characteristics. Porter and Evans (1995) report that for typical parameter values a truncation parameter of $N = 5$ is sufficient to ensure six-figure accuracy. Once the system of equations has been solved for the unknown coefficients a_n, an approximation to the reflection coefficient can be computed from (2.105). The complete solution is then readily obtained using (2.98).

It is also possible to formulate the problem as an integral equation for the jump in potential across the barrier

$$p(z) = \varphi(0^+, z) - \varphi(0^-, z), \quad z \in (-d, 0), \qquad (2.108)$$

and this leads to a problem of the type discussed in §8.1.4 (see Porter and Evans 1995 for details).

Bibliographical notes

Problems concerning the scattering of water waves by rigid barriers can be solved by numerous different techniques and are the subject of a recent book (Mandal and Chakrabarti 2000). The efficient solution procedure given above is essentially that described by Porter and Evans (1995) who also consider the scattering of waves by a bottom-mounted barrier, a barrier with a gap, a totally submerged barrier, and two identical barriers, and finally they apply the method to the calculation of the natural frequencies of oscillation of a fluid in a rectangular tank with a vertical baffle.

The radiation problem of the forced rolling motion of a thin barrier about the vertical is solved using the same technique by Evans and Porter (1996). Related Galerkin methods have been used in other contexts. For example, Lavretsky (1994) applied the method to the scattering of electromagnetic waves by a circular aperture in a rectangular wave guide.

2.5.2 A heaving truncated circular cylinder

In the vertical barrier problem considered above, the expansions in the regions to the left and to the right of the barrier were in terms of exactly the same eigenfunctions, and this leads to certain simplifications when the matching is performed. The same is true in three dimensions for the problem of wave interactions with a vertical cylindrical tube; see Thomas (1981) who solved a radiation problem for a cylindrical duct on the sea bed, fitted with a piston undergoing forced oscillations at the base. In the following example different eigenfunctions are used in the different fluid regions and the resulting equations are thus more complicated.

Consider the radiation of waves by a vertical circular cylinder occupying $r = (x^2 + y^2)^{1/2} < a$, $-d \leq z \leq 0$, making vertical oscillations in water of uniform depth $h > d$. It will be convenient to write $l = h - d$. The problem is axisymmetric so the velocity potential ϕ is a function of r and z only. The velocity potential ϕ satisfies Laplace's equation in the fluid, the free-surface condition (1.13), the bed condition (2.1), and

$$\frac{\partial \phi}{\partial r} = 0 \quad \text{on} \quad r = a, \; z \in (-d, 0), \tag{2.109}$$

$$\frac{\partial \phi}{\partial z} = U \quad \text{on} \quad r \in (0, a), \; z = -d, \tag{2.110}$$

where U is the prescribed vertical velocity. As $r \to \infty$ the potential must behave like an outgoing cylindrical wave and so in the region $r > a$ we can write (see equation 2.62)

$$\phi = Ul \sum_{n=0}^{\infty} b_n \frac{K_0(k_n r)}{K_0(k_n a)} \psi_n(z) \tag{2.111}$$

in which the $n = 0$ term corresponds to an outgoing wave (see A.5) and the factor $K_0(k_n a)$ is introduced for convenience. In the inner region, defined by $r < a$ and $z \in (-h, -d)$, the potential can be written as the sum of a particular solution which satisfies (2.110) and a general solution which satisfies a homogeneous boundary condition on the base of the cylinder. Thus we can write

$$\phi = U\phi_\mathrm{p} + Ul \sum_{n=0}^{\infty} \epsilon_n a_n \frac{I_0(\lambda_n r)}{I_0(\lambda_n a)} \cos \lambda_n (z + h), \qquad (2.112)$$

where ϵ_n is the Neumann symbol defined on page 17,

$$\phi_\mathrm{p} = \frac{1}{2l}\left((z+h)^2 - \frac{r^2}{2}\right) \qquad (2.113)$$

and $\lambda_n = n\pi/l$. The two expansions (2.111) and (2.112) must be matched by insisting that ϕ and $\partial\phi/\partial r$ are continuous across $r = a$. If the first of these conditions is imposed and the resulting equation multiplied by the set of functions $\{\cos \lambda_m (z + h); \ m = 0, 1, 2, \dots\}$ and then integrated over $(-h, -d)$, we obtain

$$g_m + a_m = \sum_{n=0}^{\infty} b_n c_{mn}, \quad m = 0, 1, 2, \dots, \qquad (2.114)$$

where

$$g_m = \frac{1}{l^2} \int_{-h}^{-d} \phi_\mathrm{p}\big|_{r=a} \cos \lambda_m (z + h)\, \mathrm{d}z = \begin{cases} \dfrac{1}{2}\left(\dfrac{1}{3} - \dfrac{a^2}{2l^2}\right) & m = 0, \\[2ex] (-1)^m / m^2 \pi^2 & m \geq 1 \end{cases} \qquad (2.115)$$

and

$$c_{mn} = \frac{1}{l} \int_{-h}^{-d} \cos \lambda_m (z + h) \psi_n(z)\, \mathrm{d}z = \frac{(-1)^m k_n \sin k_n l}{N_n l (k_n^2 - \lambda_m^2)}. \qquad (2.116)$$

It is helpful to define two new quantities

$$p_n = \frac{\lambda_n l I_0'(\lambda_n a)}{I_0(\lambda_n a)}, \qquad q_n = \frac{k_n h K_0'(k_n a)}{K_0(k_n a)} \qquad (2.117)$$

(note that $p_0 = 0$) and then continuity of $\partial\phi/\partial r$, together with (2.109), imply

$$\frac{l}{h}\sum_{n=0}^{\infty} b_n q_n \psi_n(z) = f(z), \quad z \in (-h, 0), \tag{2.118}$$

where

$$f(z) = \begin{cases} -\dfrac{a}{2l} + 2\displaystyle\sum_{n=1}^{\infty} a_n p_n \cos\lambda_n(z+h) & z \in (-h, -d), \\ 0 & z \in (-d, 0), \end{cases} \tag{2.119}$$

and then (2.118) can be multiplied by each of the functions $\psi_m(z)$ and integrated over $(-h, 0)$. This results in

$$b_m q_m = -\frac{a}{2l}c_{0m} + 2\sum_{n=1}^{\infty} a_n p_n c_{nm}, \quad m = 0, 1, 2, \dots . \tag{2.120}$$

It is possible to combine (2.114) and (2.120) by eliminating either the coefficients a_m or b_m. If one is interested in the added mass and damping coefficients, which are related to the integral of ϕ over the base of the cylinder (see equation 1.48), it is convenient to solve for the coefficients a_m. Thus we eliminate b_m and obtain the infinite system of equations

$$a_m - 2\sum_{n=1}^{\infty} a_n p_n f_{mn} = -g_m - \frac{a}{2l}f_{m0}, \quad m = 0, 1, 2, \dots , \tag{2.121}$$

where

$$f_{mn} = \sum_{s=0}^{\infty} \frac{c_{ms}c_{ns}}{q_s}. \tag{2.122}$$

Care must be taken in the numerical evaluation of this sum due to the possible occurrence of large terms resulting from the behaviour of the denominator in the definition of c_{mn} given by (2.116). In particular, the sum must be truncated well beyond $s = (h/l)\max(m, n)$.

The system of equations (2.121) can be solved numerically by truncation and Yeung (1981) reported that in order to achieve 1% accuracy, it was "rarely necessary to go beyond 20 equations". The fact that the system (2.121) converges much more rapidly than the system (2.94) for the vertical-barrier problem is a consequence of the fact that the corner at the intersection of the base and the side of the cylinder is less sharp than the corner at the tip of the barrier and hence the singularity in the derivative of ϕ there is weaker. In fact, in this case,

$$\nabla\phi = O(\rho^{-1/3}) \quad \text{as} \quad \rho = \left((r-a)^2 + (z+d)^2\right)^{1/2} \to 0. \qquad (2.123)$$

It is possible to derive a system of equations which converges even more rapidly, using a Galerkin method similar to that described in §2.5.1 for solving the vertical-barrier problem, but this will not be pursued here.

The analysis given above is only for heave oscillations, but Yeung (1981) solved all three distinct radiation problems (surge, heave, and roll) using the above technique.

Bibliographical notes

Problems concerning water-wave interaction with a truncated vertical cylinder were first studied by Miles and Gilbert (1968), and the scattering and radiation problems for the two-dimensional case of a surface-piercing rectangular block were solved by Mei and Black (1969) and Black, Mei, and Bray (1971), respectively. In these papers the appropriate eigenfunction expansions were constructed but then a variational formulation was used to obtain numerical solutions. The scattering problem with a truncated vertical cylinder was solved using a method similar to that described above by Garrett (1971). Eigenfunction expansions have also been used by Abul-Azm and Gesraha (2000) to solve the problem of oblique wave scattering by a rectangular cylinder in the free surface.

When the height of the cylinder d is zero, the problem considered above becomes that of a heaving circular disk. This problem has been solved using integral equations by MacCamy (1961) and using a systematic hierarchy of variational approximations by Miles (1987). Problems involving submerged disks have been solved using matched eigenfunction expansions by Yu and Chwang (1993) and Chwang and Wu (1994).

The matched eigenfunction expansion technique can be used to study radiation problems involving bottom-mounted circular cylinders. The surge problem, which is relevant to earthquake excitation of cylindrical storage tanks, was studied by Tung (1979), and McIver and Evans (1984b) solved the heave problem as part of an investigation into the occurrence of negative added mass (see §1.5.2).

A number of authors have used the eigenfunction expansion technique to analyze wave interaction with elliptical cylinders, in which case the solution is described in terms of Mathieu functions. The radiation and scattering of waves by a bottom-mounted surface-piercing elliptical cylinder were studied by Chen and Mei (1973) and problems concerning truncated elliptical cylinders have been solved in Stiassne and Dagan (1973) and Williams and Darwiche (1988, 1990).

2.5.3 The finite dock problem

Consider the diffraction of an oblique incident plane wave making an angle β with the positive x-axis by a rigid dock which occupies $z = 0$, $x \in [-a, a]$, $y \in (-\infty, \infty)$ in water of uniform depth h. The incident wave can be represented by the potential

$$\phi_I = e^{i\alpha(x+a)} e^{i\ell y} \psi_0(z), \tag{2.124}$$

where $\ell = k \sin \beta$ and $\alpha = k \cos \beta = (k^2 - \ell^2)^{1/2}$. From (1.8), (1.10), and (2.11), it follows that this corresponds to a wave with free surface elevation given by $\eta(x) \exp(i\ell y)$ where

$$\eta = \frac{i\omega \cosh kh}{g N_0} e^{i\alpha(x+a)}. \tag{2.125}$$

The total velocity potential for the scattering problem can similarly be written $\varphi(x, z) \exp(i\ell y)$ where $\varphi(x, z)$ satisfies the modified Helmholtz equation (2.49), the bed condition (2.1),

$$K\varphi = \varphi_z \quad \text{on } z = 0, \; |x| > a, \tag{2.126}$$

$$\varphi_z = 0 \quad \text{on } z = 0, \; |x| < a, \tag{2.127}$$

and we choose to define the reflection and transmission coefficients R and T through the far-field behaviour

$$\varphi \sim \begin{cases} \left(e^{i\alpha(x+a)} + R \, e^{-i\alpha(x+a)} \right) \psi_0(z) & \text{as } x \to -\infty, \\ T \, e^{i\alpha(x-a)} \, \psi_0(z) & \text{as } x \to \infty. \end{cases} \tag{2.128}$$

One final condition needs to be applied and that is a condition which specifies the nature of the solution near the plate edges, $(x, z) = (\pm a, 0)$. If we insist that φ is regular at these points then Wigley (1964), Theorem 3.3, implies that

$$\frac{\partial \varphi}{\partial r^{\pm}} \sim A^{\pm} \ln r^{\pm} \quad \text{as} \quad r^{\pm} = \left((x \mp a)^2 + z^2 \right)^{1/2} \to 0, \tag{2.129}$$

for some constants $A^{\pm}$.

To make the solution procedure simpler we make use of the decomposition into symmetric and antisymmetric potentials described in §2.3.3. In $x < -a$ we can expand the potentials as eigenfunction series as follows (see equation 2.50):

$$\varphi^{\pm} = \left(\mathrm{e}^{-\alpha_0(x+a)} + R^{\pm}\,\mathrm{e}^{\alpha_0(x+a)}\right)\psi_0(z) + \sum_{n=1}^{\infty} A_n^{\pm}\,\mathrm{e}^{\alpha_n(x+a)}\,\psi_n(z),$$

$$(2.130)$$

where $\alpha_0 = -\mathrm{i}\alpha = -\mathrm{i}(k^2 - \ell^2)^{1/2}$, $\alpha_n = (k_n^2 + \ell^2)^{1/2}$, $n \geq 1$, whereas for $x \in (-a, 0)$ we expand $\varphi^{\pm}$ as

$$\varphi^{\pm} = \sum_{n=0}^{\infty} \frac{\epsilon_n}{2} B_n^{\pm} \left(\mathrm{e}^{\beta_n x} \pm \mathrm{e}^{-\beta_n x}\right) \cos \lambda_n z, \qquad (2.131)$$

where $\beta_n = (\lambda_n^2 + \ell^2)^{1/2}$, $\lambda_n = n\pi/h$ and ϵ_n is the Neumann symbol defined on page 17. The edge condition (2.129) enables us to determine the behaviour of the coefficients A_n and B_n for large n. We use the result that

$$\sum_{n=1}^{\infty} n^{-1}\,\mathrm{e}^{-nx} \sim -\ln x \quad \text{as} \quad x \to 0^+ \qquad (2.132)$$

which can be derived using Mellin transforms, see e.g. Martin (1995). It follows that we must have $nA_n = O(n^{-1})$, $nB_n \exp(\beta_n a) = O(n^{-1})$ and hence that

$$A_n = O(n^{-2}), \quad B_n\,\mathrm{e}^{\beta_n a} = O(n^{-2}) \quad \text{as} \quad n \to \infty. \qquad (2.133)$$

Equations (2.130), (2.131), and (2.133) ensure that $\varphi^{\pm}$ satisfies all the appropriate equations and boundary conditions, provided $\varphi^{\pm}$ and $\partial\varphi^{\pm}/\partial x$ are continuous across $x = -a$. Note from (2.131) that there will be significant differences depending on whether or not $\ell = 0$. In particular, if $\ell = 0$ the $n = 0$ term will not contribute to φ^- or $\partial\varphi^+/\partial x$. Below, we will assume that $\ell \neq 0$ and then recover results for the $\ell = 0$ case (i.e. normal incidence) by letting $\ell \to 0$.

The continuity of $\varphi^{\pm}$ across $x = -a$ implies

$$2\psi_0(z) + \sum_{n=0}^{\infty} A_n^{\pm}\psi_n(z) = \sum_{n=0}^{\infty} \frac{\epsilon_n}{2} B_n^{\pm} \left(\mathrm{e}^{-\beta_n a} \pm \mathrm{e}^{\beta_n a}\right) \cos \lambda_n z \qquad (2.134)$$

and from the continuity of velocity,

$$\sum_{n=0}^{\infty} \alpha_n A_n^{\pm} \psi_n(z) = \sum_{n=0}^{\infty} \frac{\epsilon_n \beta_n}{2} B_n^{\pm} \left(e^{-\beta_n a} \mp e^{\beta_n a}\right) \cos \lambda_n z \qquad (2.135)$$

is obtained. Here we have written $R^{\pm} = A_0^{\pm} + 1$ for convenience. Equations (2.134) and (2.135) can be converted into an infinite system of equations in a number of different ways. For example, if (2.134) is multiplied by $\psi_m(z)$ and integrated over $(-h, 0)$, (2.135) is multiplied by $\cos \lambda_m z$ and integrated over $(-h, 0)$, and $B_m^{\pm}$ eliminated from the resulting equations we obtain

$$2\delta_{0m} + A_m^{\pm} = \sum_{n=0}^{\infty} A_n^{\pm} \left(\sum_{r=0}^{\infty} \epsilon_r c_{nr} c_{mr} \frac{\alpha_n (e^{-\beta_n a} \pm e^{\beta_n a})}{\beta_r (e^{-\beta_n a} \mp e^{\beta_n a})} \right) \qquad m \geq 0,$$
$$(2.136)$$

where

$$c_{nm} = \frac{1}{h} \int_{-h}^{0} \psi_n(z) \cos \lambda_m z \, \mathrm{d}z = \frac{k_n \sin k_n h}{h N_n (k_n^2 - \lambda_m^2)}. \qquad (2.137)$$

The system (2.136) can be solved by truncation in the usual way.

The behaviour of the unknowns A_n for large n is given by (2.133) and so the terms in the sum over n in (2.136) are $O(n^{-2})$ as $n \to \infty$. A system of equations which converges much more rapidly can be derived, however. This is because the semi-infinite dock problem (where the dock occupies $x > 0$, $z = 0$) can be solved explicitly (as will be demonstrated in Chapter 5).

If we use the orthogonality of the functions $\cos \lambda_m z$ with both (2.134) and (2.135) and then eliminate $B_m^{\pm}$ we can obtain the system

$$\sum_{n=0}^{\infty} V_n^{\pm} \left(\frac{1}{\alpha_n - \beta_m} \pm \frac{e^{-2\beta_m a}}{\alpha_n + \beta_m} \right) = \frac{1}{\alpha_0 + \beta_m} \pm \frac{e^{-2\beta_m a}}{\alpha_0 - \beta_m} \qquad m \geq 0,$$
$$(2.138)$$

where we have defined

$$V_0^{\pm} = A_0^{\pm} + 1 = R^{\pm}, \qquad (2.139)$$

$$V_n^{\pm} = \frac{A_n^{\pm} N_0 k_n \sin k_n h}{N_n k_0 \sin k_0 h}, \qquad n \geq 1. \qquad (2.140)$$

The sum over n in this system still converges like $1/n^2$, but rather than truncate this system of complex equations and solve it numerically we can take advantage of the fact that the terms $\exp(-2\beta_m a)$ all tend to zero rapidly as a/h gets large and that if the exponential terms were not present the system could be solved explicitly. The modified residue calculus technique can be used to convert the system into one which converges exponentially and it ensures that the solutions $V_n^{\pm}$ are $O(n^{-1})$ as $n \to \infty$, as required by (2.133). This procedure is described in §5.2.2 and bibliographical notes for this problem are given there.

2.6   Infinite depth

For infinite depth there is a theory equivalent to that described in the preceding sections of this chapter, although it is less frequently used. To illustrate the main ideas, the solution will be given of the wave-maker problem described for finite water depth in §2.2.1. Thus, a harmonic function $\phi(x, z)$ is sought in $x > 0$ satisfying the free-surface condition (1.13), the depth condition (1.15), the wave-maker condition (2.19), and the radiation condition (2.20) in which the wavenumber k is replaced by its deep water value $K = \omega^2/g$.

Separation of variables and application of the free-surface, depth, and radiation conditions leads to the solution

$$\phi(x, z) = A_0\, \mathrm{e}^{\mathrm{i}Kx}\, \mathrm{e}^{Kz} + \int_0^{\infty} A(\mu)\Psi_{\mu}(z)\, \mathrm{e}^{-\mu x}\, \mathrm{d}\mu \qquad (2.141)$$

which is the analogue of (2.21). Here, A_0 and $A(\mu)$ are to be found and

$$\Psi_{\mu}(z) = \mu \cos \mu z + K \sin \mu z \qquad (2.142)$$

satisfies the free-surface condition for all μ. Application of the wave-maker condition (2.19) gives

$$\left.\frac{\partial \phi}{\partial x}\right|_{x=0} = \mathrm{i}K A_0\, \mathrm{e}^{Kz} - \int_0^{\infty} A(\mu)\Psi_{\mu}(z)\mu\, \mathrm{d}\mu = U(z). \qquad (2.143)$$

Here, we assume that $U(z)$ is square integrable over $(-\infty, 0)$. The unknowns A_0 and $A(\mu)$ can be expressed in terms of the wave-maker velocity $U(z)$ by an application of the so-called 'Havelock wave-maker theorem' which states that if, for $z \in (-\infty, 0)$ and $K > 0$,

$$f(z) = C_0 \, \mathrm{e}^{Kz} + \int_0^\infty C(\mu) \Psi_\mu(z) \, \mathrm{d}\mu, \tag{2.144}$$

then

$$C_0 = 2K \int_{-\infty}^0 f(z) \, \mathrm{e}^{Kz} \, \mathrm{d}z \tag{2.145}$$

and

$$C(\mu) = \frac{2}{\pi(\mu^2 + K^2)} \int_{-\infty}^0 f(z) \Psi_\mu(z) \, \mathrm{d}z. \tag{2.146}$$

(This extension of the Fourier transform is stated without proof by Havelock 1929, but a derivation is given by Dudley 1994, example 3.7, where it is called the 'impedance transform'.) Application of this transform to (2.143) yields

$$A_0 = -2\mathrm{i} \int_{-\infty}^0 U(z) \, \mathrm{e}^{Kz} \, \mathrm{d}z, \tag{2.147}$$

$$A(\mu) = -\frac{2}{\pi\mu(\mu^2 + K^2)} \int_{-\infty}^0 U(z) \Psi_\mu(z) \, \mathrm{d}z \tag{2.148}$$

and the solution is complete. Ursell (1947, 1948) used the Havelock wave-maker theorem to obtain exact solutions to the problems of wave scattering and radiation by a thin surface-piercing vertical barrier partially immersed in an infinite-depth fluid.

Chapter 3

Multipole expansions

The eigenfunction expansion techniques described in Chapter 2 rely on the ability to expand the potential in terms of the vertical eigenfunctions constructed in §2.1. This can only be done when the fluid is made up of regions of constant finite depth and when the boundaries of all subregions coincide with coordinate lines or surfaces. In this chapter a different approach will be considered, in which the potential in the fluid is represented as a sum of singularities placed within any structures that are present. These singularities, called multipoles, are constructed to satisfy the field equation, the free-surface and bed boundary conditions (infinite and constant finite depth can be considered), and a radiation condition which demands that they behave like outgoing waves in the far field. A linear combination of these multipoles is then constructed and made to satisfy the appropriate body boundary condition. This leads to an infinite system of linear algebraic equations for the unknown coefficients in the multipole sum which can be solved numerically by truncation. Experience shows that the systems of equations that result from using a multipole method possess excellent convergence characteristics and only a few equations are needed to obtain accurate numerical answers.

The ability to apply the body boundary condition in a sensible way requires that the multipoles can be expanded in a coordinate system in which the body boundary is a coordinate surface and this restricts the class of obstacles for which the method is appropriate. The majority of wave/structure interaction problems that have been treated in this way involve circular cylinders, either horizontal or vertical, or spherical geometries, and problems of these types will be described in this chapter. The extension of the method to more complicated coordinate systems so as to treat, say, elliptical or spheroidal shapes, increases the technical difficulty but does not introduce any additional features to the method.

While the technique is perhaps best suited to single structures, multipole expansions can also be used to solve certain problems involving multiple bodies. This will be illustrated by considering a radiation problem for two submerged circular cylinders and a scattering problem for an infinite row of bottom-mounted vertical circular cylinders.

When the body boundary does not fit nicely into a particular coordinate system, it may still be possible to expand the velocity potential as a series of multipoles and use a simple collocation method to determine the unknown coefficients in the expansion. The question of the validity of the expansion for a given geometry is a complex one, however (see Athanassoulis 1984, Martin 1991b), and this approach will not be pursued here.

A systematic procedure for the construction of multipole potentials for Laplace's equation was given by Thorne (1953) and expressions from his paper together with many others which detail the properties of these functions are collected together in Appendix B. Multipoles have a different form depending on whether the singularity is in the free surface or submerged below it. As a result, problems for surface-piercing structures tend to be treated slightly differently from those concerning totally submerged structures. We will begin by illustrating the latter case and take as our example the classical problem of a submerged horizontal circular cylinder.

3.1   Isolated obstacles

3.1.1   A submerged circular cylinder

Consider the scattering of a plane wave normally-incident on an infinitely long submerged horizontal circular cylinder of radius a. The central axis of the cylinder is taken to be $x = 0$, $z = -f$ $(f > a)$ and the incident wave is taken to be from $x = -\infty$. To begin with we will assume that the water is infinitely deep and thus $\phi = \phi_\mathrm{I} + \phi_\mathrm{D}$ where (up to an arbitrary multiplicative constant)

$$\phi_\mathrm{I} = \mathrm{e}^{Kz}\,\mathrm{e}^{\mathrm{i}Kx} = \mathrm{e}^{-Kf}\sum_{n=0}^{\infty}\frac{(-Kr)^n}{n!}\,\mathrm{e}^{-\mathrm{i}n\theta}\,. \tag{3.1}$$

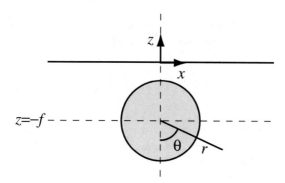

FIGURE 3.1
Coordinates for a submerged horizontal cylinder.

Here r and θ are polar coordinates defined by $z + f = -r\cos\theta$, $x = r\sin\theta$, as shown in Figure 3.1. If we wish to consider an incident wave of amplitude A, then the whole solution should be multiplied by $-igA/\omega$ (see equation 1.20). The symmetric nature of the obstacle means that we can decompose the problem into symmetric and antisymmetric parts as in §2.3.3 and for the symmetric part we write the multipole expansion

$$\phi^+ = e^{Kz}\cos Kx + \sum_{n=1}^{\infty} a^n \alpha_n \phi_n^+, \tag{3.2}$$

where the functions ϕ_n^+ are symmetric multipoles defined in §B.1 (with $\xi = 0$ and $\zeta = -f$), α_n are unknown complex constants and the factor a^n is introduced for convenience. Note that no source term is included in the summation since this would result in an instantaneous flux of fluid across the surface of the cylinder, which is physically unacceptable. The function ϕ^+ satisfies all the conditions of the problem except the structural boundary condition and in order to apply this condition, which is $\partial\phi/\partial r = 0$ on $r = a$, we must first expand ϕ^+ in terms of r and θ. From (3.1) and (B.24) we have

$$\phi^+ = e^{-Kf}\sum_{n=0}^{\infty}\frac{(-Kr)^n}{n!}\cos n\theta$$

$$+ \sum_{n=1}^{\infty} a^n \alpha_n \left(\frac{\cos n\theta}{r^n} + \sum_{m=0}^{\infty} A_{mn}^+ r^m \cos m\theta\right), \tag{3.3}$$

where A_{mn}^+ is defined by (B.25) and then application of the body boundary condition leads (because of the orthogonality of the functions $\cos n\theta$) to the system of equations

$$\alpha_m - \sum_{n=1}^{\infty} a^{m+n} A_{mn}^+ \alpha_n = \mathrm{e}^{-Kf} \frac{(-Ka)^m}{m!}, \qquad m = 1, 2, 3, \dots . \qquad (3.4)$$

This system can be solved numerically by truncating the infinite series and solving an $N \times N$ system of equations. A value of $N = 4$ will produce three decimal place accuracy unless Ka is large or the gap between the cylinder and the free surface is small.

For the antisymmetric part we write

$$\phi^- = \mathrm{i}\,\mathrm{e}^{Kz} \sin Kx + \sum_{n=1}^{\infty} a^n \beta_n \phi_n^-, \qquad (3.5)$$

where the functions ϕ_n^- are antisymmetric multipoles defined in §B.1. Application of the structural boundary condition leads (using the polar expansion of the multipoles (B.33) and the orthogonality of the functions $\sin n\theta$) to the system of equations

$$\beta_m - \sum_{n=1}^{\infty} a^{m+n} A_{mn}^- \beta_n = -\mathrm{i}\,\mathrm{e}^{-Kf} \frac{(-Ka)^m}{m!} \qquad m = 1, 2, 3, \dots, \qquad (3.6)$$

where A_{mn}^- is again defined by (B.25).

A comparison of (3.4) and (3.6) shows that the unknowns from the symmetric and antisymmetric parts of the problem are related through

$$\beta_n = -\mathrm{i}\alpha_n \qquad (3.7)$$

and so the full solution can be written

$$\phi = \phi^+ + \phi^- = \phi_{\mathrm{I}} + \sum_{n=1}^{\infty} a^n \alpha_n (\phi_n^+ - \mathrm{i}\phi_n^-). \qquad (3.8)$$

The far-field behaviour of ϕ follows from a combination of the results (B.18) for the functions ϕ_n^+ and (B.27) for the functions ϕ_n^-. This leads immediately to the remarkable conclusion that as $x \to -\infty$, $\phi \sim \phi_{\mathrm{I}}$ (i.e. the reflection coefficient is zero) and we also find that as $x \to \infty$, $\phi \sim T\phi_{\mathrm{I}}$ where the transmission coefficient is given by

$$T = 1 + 4\pi\mathrm{i}\,\mathrm{e}^{-Kf} \sum_{n=1}^{\infty} \frac{\alpha_n(-Ka)^n}{(n-1)!}. \qquad (3.9)$$

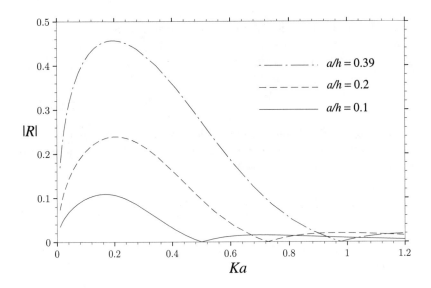

FIGURE 3.2
$|R|$ plotted against Ka for a submerged cylinder with $f/a = 1.5$
in different water depths.

It is not immediately obvious from this expression that $|T| = 1$, which
must be true for energy to be conserved (see equation 1.62). A simpler
expression for the transmission coefficient which clearly satisfies this
requirement will be derived below.

The only changes to the above analysis that are required in order to
treat the finite depth case involve replacing (3.1) by

$$\phi_{\mathrm{I}} = \cosh k(z + h)\, e^{ikx} \tag{3.10}$$

and defining A_{mn}^{+} by (B.54) and A_{mn}^{-} by (B.61). Crucially A_{mn}^{+} and A_{mn}^{-}
are no longer equal and as a result (3.7) no longer holds. Some curves of
the reflection coefficient for this case, plotted against non-dimensional
frequency, are shown in Figure 3.2. The three curves correspond to
cylinders with a constant immersion depth to radius ratio of 1.5, but in
different water depths. Whilst it is clear that R is no longer identically
zero, it is zero at certain frequencies and as one might expect, the magnitude of the reflection coefficient generally increases as the water gets
shallower.

One of the attractive features of multipole expansion methods for
this type of problem is that the evaluation of hydrodynamic forces is

straightforward. Thus for a submerged horizontal circular cylinder in infinite depth and an incident wave of amplitude A, the vertical exciting force is given, from (1.45), by

$$X_{\mathrm v} = \rho g A \int_{-\pi}^{\pi} \phi(a, \theta) \cos \theta \, a \, \mathrm{d}\theta. \tag{3.11}$$

The orthogonality of the trigonometric functions means that the only contribution to the integral comes from the terms in ϕ which are proportional to $\cos \theta$ and thus, from (3.3),

$$X_{\mathrm v} = \pi \rho g A a \left(-Ka \, \mathrm{e}^{-Kf} + \alpha_1 + \sum_{n=1}^{\infty} a^{n+1} A_{1n} \alpha_n \right). \tag{3.12}$$

This expression can be greatly simplified using (3.4) with $m = 1$ and we find that

$$X_{\mathrm v} = 2\pi \rho g A a \alpha_1. \tag{3.13}$$

A similar calculation reveals that the horizontal exciting force is given by

$$X_{\mathrm h} = -2\pi \mathrm{i} \rho g A a \alpha_1 \tag{3.14}$$

and thus that the amplitudes of these oscillatory forces are identical but that they are $90°$ out of phase. When the water depth is finite this result is no longer true and in fact the amplitude of the horizontal force (considered as a function of the non-dimensional frequency parameter Ka) has a larger maximum than in the infinite depth case and it occurs at a smaller value of Ka, whereas the maximum amplitude of the vertical force is reduced and occurs at a higher frequency (Naftzger and Chakrabarti 1979, Linton 1988).

These formulas for the exciting forces enable an expression for the transmission coefficient to be derived which is much simpler than (3.9). The amplitudes of the waves radiated to infinity by a submerged horizontal circular cylinder in heave or surge can be found from (3.13) and (3.14), respectively, if we use the Haskind relations (1.81). These amplitudes can then be substituted into (1.74) and we find that (since $R = 0$)

$$T = -\alpha_1 / \overline{\alpha_1}, \tag{3.15}$$

which clearly satisfies $|T| = 1$ as required.

Let us now consider the heave and surge radiation problems for the cylinder. The surge problem (mode 1) is antisymmetric about $x = 0$, whereas the heave problem (mode 3) is symmetric and so we can write

$$\phi_1 = \sum_{n=1}^{\infty} a^{n+1} \gamma_n^- \phi_n^-, \qquad \phi_3 = \sum_{n=1}^{\infty} a^{n+1} \gamma_n^+ \phi_n^+, \qquad (3.16)$$

for some set of unknowns $\gamma_n^\pm$. Application of the structural boundary conditions

$$\left. \frac{\partial \phi_1}{\partial r} \right|_{r=a} = \sin\theta, \qquad \left. \frac{\partial \phi_3}{\partial r} \right|_{r=a} = \cos\theta, \qquad (3.17)$$

shows that γ_m^+ satisfies the same system of equations as α_m, namely (3.4), except with $-\delta_{1m}$ on the right-hand side, whilst γ_m^- satisfies the same system of equations as β_m, (3.6), with the right-hand side again changed to $-\delta_{1m}$. These systems can again be solved by truncation and then the added mass and damping coefficients, given by (1.48), can be calculated. A similar simplification to that used to determine the exciting forces in (3.13) and (3.14) can be applied in this case and we find that

$$a_{\mu\nu}^\pm + \frac{i b_{\mu\nu}^\pm}{\omega} = \pi a^2 \rho (1 + 2\gamma_1^\pm). \qquad (3.18)$$

In infinite depth $A_{mn}^+ = A_{mn}^-$ and so the systems of equations for surge and heave are identical. We thus have $\gamma_m^+ = \gamma_m^-$ in this case and the added mass and damping coefficients for the two modes of motion are the same. Since in deep water $\phi_n^- \sim \pm i \phi_n^+$ as $x \to \pm\infty$, it follows that the amplitudes of the waves radiated to infinity, A_s and A_a, defined by (1.51) and (1.72), (1.73), satisfy

$$A_a = i A_s \qquad (3.19)$$

and the fact that $R = 0$ then follows from (1.74). Numerical results for the added mass of a submerged horizontal cylinder are given in Figure 1.2.

Scattering of obliquely incident waves can also be treated using the multipole expansion method. In this case we factor out an $\exp(i\ell y)$ dependence as shown in §2.3.2 and then require the reduced potential φ to be the solution of the modified Helmholtz equation (2.49), where, for infinite depth, ℓ is related to the incident wave angle β through

$$\alpha = (K^2 - \ell^2)^{1/2} = K \cos\beta, \qquad \ell = K \sin\beta \qquad (3.20)$$

and we clearly must have $\ell < K$. In deep water an incident wave is represented by

$$\varphi_{\mathrm{I}} = \mathrm{e}^{Kz}\,\mathrm{e}^{\mathrm{i}\alpha x} = \mathrm{e}^{-Kf} \sum_{n=0}^{\infty} \epsilon_n (-1)^n I_n(\ell r) \cos n(\theta + \mathrm{i}\gamma), \qquad (3.21)$$

where $K = \ell \cosh \gamma$, I_n is a modified Bessel function, and ϵ_n is the Neumann symbol defined on page 17 (see Gradshteyn and Ryzhik 1980, eqns 8.511(4), 8.406(3)).

The multipole expansion for φ now takes the form

$$\varphi = \varphi_{\mathrm{I}} + \sum_{n=0}^{\infty} \alpha_n \varphi_n^+ + \sum_{n=1}^{\infty} \beta_n \varphi_n^-, \qquad (3.22)$$

where φ_n^+ and φ_n^- are, respectively, the symmetric and antisymmetric multipoles defined in §B.5. We note that, unlike in (3.2), the $n = 0$ term is included in the sum of symmetric multipoles (the completeness of multipole expansions for oblique wave problems is discussed in Ursell 1968).

The boundary condition on the structure then shows that the unknown coefficients α_n and β_n must satisfy the systems of equations

$$\alpha_m + F_m \sum_{n=0}^{\infty} A_{mn}^+ \alpha_n = -\epsilon_m \, \mathrm{e}^{-Kf} (-1)^m F_m \cosh m\gamma \quad m = 0, 1, 2, \dots,$$
$$(3.23)$$

where $F_m = I_m'(\ell a)/K_m'(\ell a)$ and A_{mn}^+ is defined by (B.116), and

$$\beta_m + F_m \sum_{n=1}^{\infty} A_{mn}^- \beta_n = 2\mathrm{i}\,\mathrm{e}^{-Kf} (-1)^m F_m \sinh m\gamma \quad m = 1, 2, 3, \dots,$$
$$(3.24)$$

where A_{mn}^- is defined by (B.121). These equations can be solved numerically by truncation and again the convergence characteristics are excellent. Just as for normal incidence, the truncation required for a given accuracy increases as the submergence decreases or as Ka increases. In this case we also find that the truncation size must be increased as $\beta \to \pi/2$. The reflection and transmission coefficients can be determined from

$$R = 2\pi K \alpha^{-1} e^{-Kf} \sum_{n=0}^{\infty} (-1)^n (i\alpha_n \cosh n\gamma + \beta_n \sinh n\gamma), \qquad (3.25)$$

$$T = 1 + 2\pi K \alpha^{-1} e^{-Kf} \sum_{n=0}^{\infty} (-1)^n (i\alpha_n \cosh n\gamma - \beta_n \sinh n\gamma), \qquad (3.26)$$

which follow from (3.22) and the far-field form of the multipoles (B.112), (B.118). Approximate expressions for R and T based on an integral equation formulation were derived by Levine (1965), who established that the reflection coefficient is not identically zero for the oblique incidence case. The extension of this scattering problem to finite depth is straightforward; the multipoles φ_n^+ and φ_n^- must simply be replaced by their finite depth counterparts given in §B.6.

A further extension is to consider the case $\ell > K$. With this condition it is no longer possible for propagating waves to exist as $|x| \to \infty$. Nevertheless, for fixed ℓ, one can look for values of K in the interval $(0, \ell)$ at which non-trivial solutions to the boundary-value problem exist; such solutions corresponding to waves propagating along the cylinder and decaying exponentially away from the cylinder. Solutions of this type are known as trapped modes. Mathematically, we need to solve (3.23) or (3.24) with zero on the right-hand side. Note that because $\ell > K$, all the terms in these equations are real. The problem boils down to finding values of K for which either of the infinite determinants $|\delta_{mn} + F_m A_{mn}^{\pm}|$ vanishes; if the plus sign is taken then the resulting modes are symmetric about $x = 0$ whereas they are antisymmetric about this line if the minus sign is used.

Bibliographical notes

The fact that the reflection coefficient for a submerged horizontal circular cylinder in infinite depth water is zero, for all incident wave frequencies, was discovered by Dean (1948) using a conformal mapping technique and the multipole expansion method described above was presented by Ursell (1950a), who also proved (Ursell 1950b) that the systems (3.4) and (3.6) always possess unique solutions. The first person to consider the forces on such a structure was Ogilvie (1963) and the calculation of the hydrodynamic characteristics via the heave and surge radiation problems and the reciprocity relations can be found in Evans et al. (1979), who investigated the possibility of using a submerged cylinder as a wave energy device. More recently, Linton and McIver (1995) have shown that zero reflection occurs at all frequencies when a circular cylinder is submerged in the lower layer of an infinitely deep two-layer fluid. In that case there are two possible wavenumbers for a given frequency, one associated with the free surface and one with the interface, but remarkably the reflection coefficient vanishes identically whichever incident wavenumber is chosen.

Multipole expansions have been used to calculate the hydrodynamic characteristics of many other submerged bodies. Circular cylinders in finite depth were treated in Evans and Linton (1989), who examined the possibility of using a tethered submerged cylinder as a breakwater. The case of a submerged sphere in deep water can be found in Srokosz (1979) and Wang (1986) and the finite depth case was treated by Linton (1991). The heave added mass and damping coefficients of a submerged torus in deep water, relevant to the design of ring-hulled semi-submersible oil platforms, were computed by Hulme (1985).

Ursell (1951) proved that provided ka was small enough, values of K at which trapped modes can occur above a circular cylinder do exist for the symmetric problem and Jones (1953) provided a more general proof which removed the restriction on the size of cylinder. Numerical computations of the symmetric trapped-mode frequencies were performed by McIver and Evans (1985). Much less work has been done on the antisymmetric problem, but such modes have been found numerically (Martin 1989).

The method of multipoles may be applied to many problems in different physical contexts. For example, Linton (1995) applied the method to the problems of potential flow and Stokes (viscous) flow past, and acoustic scattering by, a sphere on the axis of a circular cylindrical tube.

3.1.2   A heaving hemisphere

The use of multipole expansions is not restricted to the study of wave interaction with submerged bodies; problems involving surface-piercing structures can also be treated. To illustrate this we describe the method used to solve the problem of a floating hemisphere undergoing forced vertical oscillations in deep water. Spherical polar coordinates (r, θ, α) and cylindrical polar coordinates (R, α, z) are used, defined by

$$x = R \cos \alpha, \quad y = R \sin \alpha, \quad z = -r \cos \theta, \quad R = r \sin \theta. \qquad (3.27)$$

The problem is clearly axisymmetric and so the solution ϕ will be independent of α.

For this problem it is most convenient to write ϕ as the sum of a source term Ψ, given by (B.71), and a series of wave-free potentials ψ_{2n}, given by (B.84) with $m = 0$. A source term is required as there is a non-zero instantaneous net flux of fluid across the mean position of the submerged surface of the hemisphere. The amplitude of the waves radiated to infinity is determined entirely by the coefficient of the source term in this expansion. (Since the source is axisymmetric, this expansion is clearly only applicable to axisymmetric problems like this one; if one were solving a different radiation problem one would have to include— as Hulme (1982) did for the surge problem—a higher-order multipole with the appropriate wave-like behaviour at infinity.) Thus we write the multipole expansion as

$$\phi = c_0 \left(a^2 \Psi + \sum_{n=1}^{\infty} a^{2n+2} c_n \psi_{2n} \right), \qquad (3.28)$$

where we tacitly assume that the coefficient multiplying the source term is non-zero (Hulme's calculations reveal this to be always the case) and the body boundary condition is

$$\frac{\partial \phi}{\partial r} = \cos\theta \quad \text{on} \quad r = a \qquad (3.29)$$

from which, for $\theta \in (0, \pi/2)$,

$$F(\mu, Ka) - \sum_{n=1}^{\infty} c_n \left(Ka P_{2n-1}(\mu) + (2n+1) P_{2n}(\mu) \right) = \frac{P_1(\mu)}{c_0}, \qquad (3.30)$$

where we have written $\mu = \cos\theta$ and

$$F(\mu, Ka) = a^2 \frac{\partial \Psi}{\partial r} \bigg|_{r=a}. \qquad (3.31)$$

If (3.30) is integrated with respect to μ over $(0, 1)$ we obtain

$$\int_0^1 F(\mu, Ka) \, d\mu - Ka \sum_{n=1}^{\infty} c_n I_{0,2n-1} = \frac{I_{0,1}}{c_0} = \frac{1}{2c_0}, \qquad (3.32)$$

where

$$I_{m,n} = \int_0^1 P_m(\mu) P_n(\mu) \, d\mu, \qquad (3.33)$$

integrals which can be determined in closed form in terms of elementary functions (see Hulme 1982).

The expression for c_0 given by (3.32) is now substituted back into (3.30) to give

$$\sum_{n=1}^{\infty} c_n \left((2n+1)P_{2n}(\mu) + Ka[P_{2n-1}(\mu) - 2P_1(\mu)I_{0,2n-1}]\right) = G(\mu, Ka),$$
(3.34)

where

$$G(\mu, Ka) = F(\mu, Ka) - 2P_1(\mu) \int_0^1 F(\nu, Ka)\, d\nu. \qquad (3.35)$$

To convert (3.34) into an infinite system of equations we multiply by each element of the complete set $\{P_{2m}(\mu);\ m = 1, 2, 3, \dots\}$ in turn and integrate over $(0, 1)$. We find that the unknowns c_n, $n = 1, 2, 3, \dots$, must satisfy the infinite system of equations

$$\frac{2m+1}{4m+1}c_m + Ka\sum_{n=1}^{\infty} A_{mn}c_n = \beta_m, \quad m = 1, 2, 3, \dots, \qquad (3.36)$$

where

$$A_{mn} = I_{2m,2n-1} - 2I_{2m,1}I_{0,2n-1}, \qquad (3.37)$$

$$\beta_m = J_{2m} - 2J_0 I_{2m,1}, \qquad (3.38)$$

$$J_m = \int_0^1 F(\mu, Ka)P_m(\mu)\, d\mu \qquad (3.39)$$

and then c_0 can be determined from (3.32). With the aid of the source expansion (B.74) we can show that

$$J_m = -I_{m,0} - Ka\sum_{n=1}^{\infty} \frac{(-Ka)^n}{(n-1)!} \left.\frac{\partial I_{m,\nu}}{\partial \nu}\right|_{\nu=n}$$

$$+ Ka\sum_{n=0}^{\infty} \frac{(-Ka)^n}{n!}\left(n[\psi(n) + \pi i - \ln Ka] - 1\right) I_{m,n}, \qquad (3.40)$$

where $\psi(n) = -\gamma + \sum_{s=1}^{n-1} s^{-1}$ is the Digamma function and γ is Euler's constant.

The behaviour of the coefficients c_n in the limit as $Ka \to 0$ can easily be determined since it follows from (3.36) that

$$c_n = \frac{4n+1}{2n+1} \beta_n [1 + O(Ka)] \quad \text{as} \quad Ka \to 0. \tag{3.41}$$

The asymptotic form for β_n in this limit can also be found and it can then be shown that the added mass and damping coefficients (defined in equation 1.48) satisfy

$$M^{-1} a_{33} = L - \tfrac{3}{4} Ka \ln Ka + O(Ka), \tag{3.42}$$

$$(M\omega)^{-1} b_{33} = \tfrac{3}{4} \pi Ka + O((Ka)^2), \tag{3.43}$$

where

$$L = 3 \sum_{n=0}^{\infty} \frac{4n+1}{2n+1} I_{2n,1}^2 \approx 0.83095 \tag{3.44}$$

and M is the mass of fluid displaced by the hemisphere.

Bibliographical notes

The method described above was originally implemented by Havelock (1955) and improved on by Hulme (1982), who also solved the surge problem using the same technique. Havelock's work was an extension to three dimensions of the pioneering work of Ursell (1949), who introduced the multipole expansion method when analyzing wave interactions with a floating semicircular cylinder in infinite depth. The corresponding finite depth problem was solved by Yu and Ursell (1961).

The behaviour of the added mass for a floating hemisphere at low frequency, as given by (3.42), was derived using an entirely different method by Kotik and Mangulis (1962). (They did not calculate the constant L, but showed that for an arbitrary three-dimensional body in deep water whose intersection with the free surface has area πa^2, the added mass in heave satisfies $a_{33}(Ka) - a_{33}(0) \sim -\tfrac{1}{2} \rho \pi a^3 Ka \ln Ka$ as $Ka \to 0$.) This result, which is in agreement with the calculations of Havelock (1955), contradicts the calculations of Barakat (1962), who claimed that the added mass for a heaving hemisphere is initially a decreasing function of Ka as Ka increases from zero.

3.1.3 Sloshing in a hemisphere

A technique akin to multipole expansions can also be used to find the natural frequencies of oscillation of certain fluid-filled containers.

Following on from the previous example we consider the case of a half-full sphere. Thus we seek a function ϕ, harmonic in $r < a$, $\theta \in [0, \pi/2)$, $\alpha \in [0, 2\pi)$, which satisfies $\partial\phi/\partial r = 0$ on $r = a$ as well as the free-surface boundary condition, which in spherical polar coordinates takes the form

$$K\phi = \frac{1}{r}\frac{\partial\phi}{\partial\theta} \quad \text{on} \quad \theta = \frac{\pi}{2}. \tag{3.45}$$

The natural frequencies of oscillation then correspond to those values of K for which non-trivial solutions can be found.

If we assume that the velocity potential has a $\cos M\alpha$ dependence then it follows that it can be expanded in spherical harmonics as

$$\phi(r, \theta, \alpha) = \cos M\alpha \sum_{n=M}^{\infty} c_n r^n P_n^M(\mu), \tag{3.46}$$

where we have written $\mu = \cos\theta$. If the free-surface boundary condition is applied we find, since $P_n^M(0) = 0$ if $n + M$ is odd and

$$P_n^{M\prime}(0) = (n + M)P_{n-1}^M(0), \tag{3.47}$$

that ϕ must be of the form

$$\phi = \cos M\alpha \sum_{n=m}^{\infty} c_{2n+1}\left(r^{2n+1}P_{2n+1}^M(\mu) - (2n + M + 1)\frac{r^{2n}}{K}P_{2n}^M(\mu)\right) \tag{3.48}$$

if $M = 2m$ is even, whereas

$$\phi = \cos M\alpha \sum_{n=m+1}^{\infty} c_{2n}\left(r^{2n}P_{2n}^M(\mu) - (2n + M)\frac{r^{2n-1}}{K}P_{2n-1}^M(\mu)\right) \tag{3.49}$$

if $M = 2m+1$ is odd. Thus for this interior problem we have constructed combinations of regular solutions of Laplace's equation which satisfy the free-surface boundary condition and which are the analogues of the singular solutions (B.84) and (B.85) for the exterior problem.

If the body boundary condition is now applied and the orthogonality result, valid if $p + q$ is even,

$$I_{p,q}^M \equiv \int_0^1 P_p^M(\mu)P_q^M(\mu)\,\mathrm{d}\mu = \frac{\delta_{pq}(q + M)!}{(2q + 1)(q - M)!}, \tag{3.50}$$

used, we find that (except when $M = 0$) the values of $(Ka)^{-1}$ which give rise to non-trivial solutions are the eigenvalues of an infinite matrix $\mathbf{A}$ whose elements are given by

$$A_{ij} = \frac{(4i + 2M - 3)(2j + M - 1)(2i - 2)!}{(2i + M - 2)(2i + 2M - 1)!} I^M_{2j+M-1,2i+M-2}. \quad (3.51)$$

When $M = 0$ a complication arises because the summation in (3.48) now starts from zero and thus contains a constant term, which vanishes when the body boundary condition is applied. However, we can in this case solve for c_1 in terms of c_{2n+1}, $n = 1, 2, 3, \ldots$, and show that $(Ka)^{-1}$ must be an eigenvalue of

$$A_{ij} = \frac{4i + 1}{2i} \left(I_{2j+1,2i} - 2 I_{2j+1,0} I_{1,2i} \right). \quad (3.52)$$

The integrals $I^M_{p,q}$ are given in terms of elementary functions in Hulme (1982) and so the sloshing frequencies can easily be computed numerically by truncating $\mathbf{A}$ to an $N \times N$ matrix and increasing N until the desired accuracy is achieved. The more eigenvalues that are required the larger the value of N that is needed, but in practice this technique is only sensible for determining the first couple of sloshing frequencies because asymptotic expressions can be derived for the higher modes. Thus for the case considered here Davis (1975) showed that for $m \geq 1$ the nth eigenvalue is given for large n by the asymptotic expression

$$K_n a \sim j'_{mn} - \frac{1}{4 j'_{mn}} - \frac{1}{3\pi j'^2_{mn}} - \frac{1}{2 j'^3_{mn}} \left(m^3 + \frac{661}{768} + \frac{4}{3\pi^2} \right), \quad (3.53)$$

where j'_{mn} is the nth zero of $J'_m(x)$. For $m = 0$, the numbers j'_{mn} should be replaced by $j'_{m,n+1}$ because the first zero of $J'_0(x)$ is zero. Some numerical results are shown in Table 3.1 for the first three azimuthal modes $m = 0, 1, 2$. The table shows results computed from (3.51) and (3.52) using an 8×8 and a 16×16 truncation, together with the correct value to three decimal places and the value computed from (3.53). The high accuracy of the asymptotic formula, for all but the lowest mode, is clear.

The method described here was applied to the two-dimensional problem of sloshing in a cylinder of semi-circular cross-section and the three-dimensional problem of sloshing in a semi-circular horizontal cylinder of finite length (as well as to the hemisphere problem) in Evans and Linton (1993b).

	$m = 0$		$m = 1$		$m = 2$	
8×8	3.744	6.896	1.560	5.272	2.819	6.622
16×16	3.745	6.976	1.560	5.275	2.820	6.659
correct to 3dp.	3.745	6.976	1.560	5.276	2.820	6.659
asymptotic	3.750	6.976	1.514	5.274	2.873	6.658

Table 3.1 Sloshing frequencies Ka for a half-full sphere.

3.2 Multiple bodies

3.2.1 Two submerged circular cylinders

It is possible to use the multipole method to solve problems involving the interaction of waves with more than one obstacle of the same type. The general procedure involves writing the solution as a sum of multipole expansions corresponding to each structure and then re-expanding each of these expansions about the centre of each of the other structures so that the boundary conditions can be applied. To illustrate the method we will consider a relatively simple case, namely two identical submerged circular cylinders in heave motion in deep water.

The central axes of the cylinders, which both have radius a, are taken to be $(x, z) = (\pm b, -f)$ so that the resulting solution is symmetric about $x = 0$. We use multipoles $\phi_n^{(1)\pm}$ singular at $x = b$ $z = -f$, where a plus sign signifies a multipole symmetric about $x = b$ and a minus sign indicates a multipole antisymmetric about this line (such potentials are defined in §B.1). Similarly for the multipoles $\phi_n^{(2)\pm}$, which are singular at $x = -b$, $z = -f$. Then we can consider the potential

$$\phi = \sum_{n=1}^{\infty} a^{n+1} \left[\gamma_n^+ \left(\phi_n^{(1)+} + \phi_n^{(2)+} \right) + \gamma_n^- \left(\phi_n^{(1)-} - \phi_n^{(2)-} \right) \right], \qquad (3.54)$$

for some set of unknowns $\gamma_n^\pm$. This potential is symmetric about the line $x = 0$ by construction and so we only need to apply the body boundary condition on the surface of one of the cylinders; the other will be satisfied automatically.

The multipoles $\phi_n^{(1)\pm}$ can be expanded in terms of polar coordinates (r_1, θ_1) centred at $(b, -f)$ using (B.24). (Note that r_1 and θ_1 have a different meaning in Appendix B.) For the multipole $\phi_n^{(2)+}$ we have

$$\phi_n^{(2)+} = \frac{\cos n\theta_2}{r_2^n} + \frac{(-1)^n}{(n-1)!} \int_0^\infty \frac{\mu+K}{\mu-K} \mu^{n-1} e^{\mu(z-f)} \cos\mu(x+b) \, d\mu,$$
$$(3.55)$$

where (r_2, θ_2) are polar coordinates centred at $(-b, -f)$. The integral is readily expanded in terms of (r_1, θ_1) by writing

$$\cos\mu(x+b) = \cos\mu(x-b+2b)$$
$$= \cos\mu(x-b)\cos 2\mu b - \sin\mu(x-b)\sin 2\mu b \quad (3.56)$$

so that, for $r < 2f$,

$$\phi_n^{(2)+} = \frac{\cos n\theta_2}{r_2^n} + \sum_{m=0}^\infty P_{mn}^+ r_1^m \cos m\theta_1 + \sum_{m=1}^\infty Q_{mn}^+ r_1^m \sin m\theta_1, \quad (3.57)$$

where

$$P_{mn}^+ = \frac{(-1)^{m+n}}{m!(n-1)!} \int_0^\infty \frac{\mu+K}{\mu-K} \mu^{m+n-1} e^{-2\mu f} \cos 2\mu b \, d\mu, \quad (3.58)$$

$$Q_{mn}^+ = \frac{(-1)^{m+n}}{m!(n-1)!} \int_0^\infty \frac{\mu+K}{\mu-K} \mu^{m+n-1} e^{-2\mu f} \sin 2\mu b \, d\mu. \quad (3.59)$$

It remains to express $r_2^{-n} \cos n\theta_2$ in terms of r_1 and θ_1. This is facilitated by the introduction of a second complex unit j (independent of the complex unit i introduced when the time-dependence was factored out in equation 1.10) so that any point in the (x, z)-plane can be represented by a complex number $w = x + jz$. The central axes of the two cylinders are located at $w_1 = b - jf$ and $w_2 = -b - jf$ and clearly $w_1 - w_2 = 2b$. For an arbitrary point w we have

$$w = w_1 + r_1 e^{j(\theta_1 - \pi/2)} = w_2 + r_2 e^{j(\theta_2 - \pi/2)}. \quad (3.60)$$

Consider the complex function

$$f_n(w) = \frac{(-j)^n}{(w-w_2)^n} = \frac{e^{-jn\theta_2}}{r_2^n}. \quad (3.61)$$

This function is analytic everywhere in the complex j-plane except at $w = w_2$ and so it can be expanded as a power series about $w = w_1$ and the expansion will be valid on $r_1 = a$. Thus

$$f_n(w) = \sum_{m=0}^{\infty} \frac{d^m f_n}{dw^m}\bigg|_{w=w_1} \frac{(w - w_1)^m}{m!} \tag{3.62}$$

$$= \sum_{m=0}^{\infty} \frac{(-1)^m (n + m - 1)!}{m!(n-1)!(2b)^{n+m}} r_1^m \, e^{j[m\theta_1 - (n+m)\pi/2]} \,. \tag{3.63}$$

It follows that

$$\frac{\cos n\theta_2}{r_2^n} = \sum_{m=0}^{\infty} \frac{(-1)^m (n + m - 1)!}{m!(n-1)!(2b)^{n+m}} r_1^m \cos[m\theta_1 - (n + m)\pi/2]. \tag{3.64}$$

This expression can then be substituted into (3.57) to provide an expansion of $\phi_n^{(2)+}$ in terms of r_1 and θ_1, valid on $r_1 = a$.

For $\phi_n^{(2)-}$ we obtain

$$\phi_n^{(2)-} = \frac{\sin n\theta_2}{r_2^n} + \sum_{m=0}^{\infty} P_{mn}^- r_1^m \cos m\theta_1 + \sum_{m=1}^{\infty} Q_{mn}^- r_1^m \sin m\theta_1, \tag{3.65}$$

where

$$P_{mn}^- = \frac{(-1)^{m+n+1}}{m!(n-1)!} \int_0^{\infty} \frac{\mu + K}{\mu - K} \mu^{m+n-1} e^{-2\mu f} \sin 2\mu b \, d\mu, \tag{3.66}$$

$$Q_{mn}^- = \frac{(-1)^{m+n}}{m!(n-1)!} \int_0^{\infty} \frac{\mu + K}{\mu - K} \mu^{m+n-1} e^{-2\mu f} \cos 2\mu b \, d\mu. \tag{3.67}$$

and

$$\frac{\sin n\theta_2}{r_2^n} = - \sum_{m=0}^{\infty} \frac{(-1)^m (n + m - 1)!}{m!(n-1)!(2b)^{n+m}} r_1^m \sin[m\theta_1 - (n + m)\pi/2]. \tag{3.68}$$

Equations (3.57) and (3.65), together with (3.64) and (3.68), can then be substituted into (3.54) to provide an expansion for ϕ solely in terms of the polar coordinates r_1 and θ_1 of the form

$$\phi = \sum_{n=1}^{\infty} a^{n+1} \Bigg[\gamma_n^+ \left(\frac{\cos n\theta_1}{r_1^n} + \sum_{m=0}^{\infty} r_1^m [C_{mn}^+ \cos m\theta_1 + S_{mn}^+ \sin m\theta_1] \right)$$

$$+ \gamma_n^- \left(\frac{\sin n\theta_1}{r_1^n} + \sum_{m=0}^{\infty} r_1^m [C_{mn}^- \cos m\theta_1 + S_{mn}^- \sin m\theta_1] \right) \Bigg], \tag{3.69}$$

where the coefficients $C_{mn}^{\pm}$ and $S_{mn}^{\pm}$ are known. This expansion is valid on $r_1 = a$ and so allows the body boundary condition, $\partial\phi/\partial r_1 = 0$ on $r_1 = a$, to be applied.

This results in the coupled systems of equations

$$\gamma_m^+ - \sum_{n=1}^{\infty} a^{m+n}(C_{mn}^+ \gamma_n^+ + C_{mn}^- \gamma_n^-) = -\delta_{1m}, \qquad (3.70)$$

$$\gamma_m^- - \sum_{n=1}^{\infty} a^{m+n}(S_{mn}^+ \gamma_n^+ + S_{mn}^- \gamma_n^-) = 0, \qquad (3.71)$$

$m = 1, 2, 3, \ldots$ in both cases, which can be solved numerically by truncation. The extension to finite depth is straightforward; the multipoles of §B.1 are simply replaced by their equivalents from §B.2.

Bibliographical notes

The heave problem for two identical half-immersed cylinders in deep water was solved by Wang and Wahab (1971) and both the heave and surge problem for two identical submerged cylinders in deep water were solved by Wang (1981). All these problems have relevance to the design of catamaran-type vessels. The general problem of wave radiation and scattering by a group of horizontal cylinders with different radii and immersion depths in infinitely deep water was solved by O'Leary (1985). The case of two cylinders moving in surge but exactly out of phase with each other, which is equivalent to a surging cylinder next to a vertical wall, was treated by Linton (1988) for the case of finite depth.

The three-dimensional radiation and scattering problems for a group of submerged spheres were solved using multipole expansions by Wu (1995).

3.2.2 A row of vertical circular cylinders

Consider a row of identical bottom-mounted vertical circular cylinders of radius a arranged so that the centres of the cylinder cross-sections are at $(x, y) = (0, 2md)$, $m = 0, \pm 1, \pm 2, \ldots$. Define horizontal polar coordinates centred on the origin by $x = r\cos\theta$, $y = r\sin\theta$ in the usual way. We assume that a plane wave making an angle β with the positive x-axis is incident on the cylinders and the depth dependence of the problem is factored out as described in §2.4 so that the problem is reduced to one in the (x, y)-plane and we look for solutions to the

Helmholtz equation (2.74). The incident wave is characterized by

$$\varphi_{\mathrm{I}} = \mathrm{e}^{\mathrm{i}\alpha x + \mathrm{i}\ell y} = \mathrm{e}^{\mathrm{i}kr\cos(\theta-\beta)} = \sum_{m=0}^{\infty} \epsilon_m \mathrm{i}^m J_m(kr) \cos m(\theta - \beta) \quad (3.72)$$

(Gradshteyn and Ryzhik 1980, eqn 8.511(4)), where J_m is a Bessel function of the first kind,

$$\alpha = k\cos\beta, \qquad \ell = k\sin\beta, \quad (3.73)$$

and ϵ_m is the Neumann symbol defined on page 17.

Since the incident wave is periodic in the y-direction and the array of cylinders extends over the whole y-axis, we can seek a scattered wave field φ which has the term $\exp(\mathrm{i}\ell y)$ in common with the incident wave. This, together with the periodicity of the geometry, implies that

$$\varphi(x, y) = \mathrm{e}^{\mathrm{i}\ell y} \, \psi(x, y), \quad (3.74)$$

where ψ is periodic in y with period $2d$. It follows that we need only consider the strip $|y| < d$ and from (3.74) we can derive the two independent periodicity conditions

$$\varphi|_{y=d} = \mathrm{e}^{2\mathrm{i}\ell d} \, \varphi|_{y=-d} \quad (3.75)$$

and

$$\left.\frac{\partial\varphi}{\partial y}\right|_{y=d} = \mathrm{e}^{2\mathrm{i}\ell d} \left.\frac{\partial\varphi}{\partial y}\right|_{y=-d}. \quad (3.76)$$

The approach we now take is to construct multipoles φ_n^+, φ_n^-, symmetric and antisymmetric, respectively, about the line $x = 0$. Such functions satisfy the Helmholtz equation in the strip $|y| < d$, $x \in (-\infty, \infty)$, except at the origin where they are singular, and the periodicity conditions (3.75), (3.76). The derivation of these functions for $\ell = 0$ is described in detail in Linton and Evans (1992) and for non-zero ℓ they are listed together with some of their important properties in Linton and Evans (1993c).

We thus express the velocity potential as

$$\varphi = \varphi_{\mathrm{I}} + \sum_{n=0}^{\infty} a_n \varphi_n^+ + \sum_{n=1}^{\infty} b_n \varphi_n^-, \quad (3.77)$$

where the multipoles are defined by

$$\varphi_{2n}^+(x,y) = -\frac{2i}{\pi}(-1)^n \int_0^\infty \frac{f(y,t)\cos kxt\, c_{2n}(t)}{\gamma(t)\Delta(t)}\,dt, \qquad (3.78)$$

$$\varphi_{2n+1}^+(x,y) = \frac{2}{\pi}\,\mathrm{sgn}(y)(-1)^n \int_0^\infty \frac{g(y,t)\cos kxt\, c_{2n+1}(t)}{\gamma(t)\Delta(t)}\,dt, \qquad (3.79)$$

$$\varphi_{2n}^-(x,y) = -\frac{2}{\pi}\,\mathrm{sgn}(y)(-1)^n \int_0^\infty \frac{g(y,t)\sin kxt\, s_{2n}(t)}{\gamma(t)\Delta(t)}\,dt, \qquad (3.80)$$

$$\varphi_{2n+1}^-(x,y) = -\frac{2i}{\pi}(-1)^n \int_0^\infty \frac{f(y,t)\sin kxt\, s_{2n+1}(t)}{\gamma(t)\Delta(t)}\,dt, \qquad (3.81)$$

with

$$f(y,t) = e^{2i\ell d\,\mathrm{sgn}(y)}\sinh k\gamma(t)|y| + \sinh[k\gamma(t)(2d-|y|)], \qquad (3.82)$$

$$g(y,t) = e^{2i\ell d\,\mathrm{sgn}(y)}\cosh k\gamma(t)y - \cosh[k\gamma(t)(2d-|y|)], \qquad (3.83)$$

$$\Delta(t) = \cosh 2k\gamma(t)d - \cos 2\ell d \qquad (3.84)$$

and

$$\gamma(t) = -i(1-t^2)^{1/2} = (t^2-1)^{1/2}, \qquad (3.85)$$

$$c_m(t) = \cos(m\sin^{-1}t) = \cos[m(\pi/2 + i\cosh^{-1}t)], \qquad (3.86)$$

$$s_m(t) = \sin(m\sin^{-1}t) = \sin[m(\pi/2 + i\cosh^{-1}t)]. \qquad (3.87)$$

The first and second expressions in the last three of these definitions correspond to $t \leq 1$ and $t > 1$, respectively.

The unknown constants a_n, b_n, can then be determined by applying the boundary condition on the cylinder, which is $\partial\varphi/\partial r = 0$ on $r = a$. Using the appropriate polar coordinate expansions of the multipoles (see Linton and Evans 1993c), we can write, for $r < 2d$,

$$\varphi(r,\theta) = \sum_{m=0}^\infty C_m(r)\cos m\theta' + \sum_{m=1}^\infty S_m(r)\sin m\theta', \qquad (3.88)$$

where $\theta' = \pi/2 - \theta$ and, with $\beta' = \pi/2 - \beta$,

$$C_m(r) = \epsilon_m i^m J_m(kr)\cos m\beta' + a_m H_m^{(1)}(kr) + J_m(kr)\sum_{n=0}^\infty a_n E_{mn}^+, \qquad (3.89)$$

$$S_m(r) = 2i^m J_m(kr)\sin m\beta' + b_m H_m^{(1)}(kr) + J_m(kr)\sum_{n=1}^\infty b_n E_{mn}^-. \qquad (3.90)$$

Here

$$
E_{mn}^{+} =
\begin{cases}
\dfrac{2\epsilon_m}{\pi} i^{n-m+1} \displaystyle\int_0^\infty \dfrac{e^{-2k\gamma d} - \cos 2\ell d}{\gamma \Delta}\, c_n c_m\, \mathrm{d}t, & n+m \text{ even,} \\[4mm]
\dfrac{2\epsilon_m}{\pi} i^{n-m} \displaystyle\int_0^\infty \dfrac{\sin 2\ell d}{\gamma \Delta}\, c_n c_m\, \mathrm{d}t, & n+m \text{ odd,}
\end{cases}
$$

$$\tag{3.91}$$

$$
E_{mn}^{-} =
\begin{cases}
\dfrac{4}{\pi} i^{n-m+1} \displaystyle\int_0^\infty \dfrac{e^{-2k\gamma d} - \cos 2\ell d}{\gamma \Delta}\, s_n s_m\, \mathrm{d}t, & n+m \text{ even,} \\[4mm]
\dfrac{4}{\pi} i^{n-m} \displaystyle\int_0^\infty \dfrac{\sin 2\ell d}{\gamma \Delta}\, s_n s_m\, \mathrm{d}t, & n+m \text{ odd.}
\end{cases}
$$

$$\tag{3.92}$$

The body boundary condition implies that $C_m'(a) = S_m'(a) = 0$, from which

$$
a_m + Z_m \sum_{n=0}^{\infty} a_n E_{mn}^{+} = -\epsilon_m i^m Z_m \cos m\beta', \qquad m = 0, 1, \ldots, \tag{3.93}
$$

$$
b_m + Z_m \sum_{n=1}^{\infty} b_n E_{mn}^{-} = -2i^m Z_m \sin m\beta', \qquad m = 1, 2, \ldots, \tag{3.94}
$$

where $Z_m = J_m'(ka)/H_m^{(1)'}(ka)$. Thus the a_n's and the b_n's each satisfy an infinite system of linear algebraic equations which can be solved efficiently by truncation. The fact that the equations decouple is a consequence of the symmetry of the geometry about $x = 0$. The systems of equations (3.93) and (3.94) can be used to simplify the polar coordinate expansion (3.88), resulting in

$$
\varphi(r, \theta) = \sum_{m=0}^{\infty} \left(H_m^{(1)}(kr) - Z_m^{-1} J_m(kr) \right) (a_m \cos m\theta' + b_m \sin m\theta')
$$

$$\tag{3.95}$$

valid for $r < 2d$, which takes a particularly simple form on $r = a$ due to the Wronskian relation (A.7).

In order to determine the reflected and transmitted wave fields, the far-field forms of the multipoles $\varphi_n^{\pm}$ are needed. These are related to the positive real zeros of the function $\Delta(t)$ which appears in the denominators of the multipoles, i.e. $t = t_p$, $p = -\mu, \ldots, \nu$, where

$$
t_p = (1 - (\ell_p/k)^2)^{1/2}, \qquad \ell_p = \ell + p\pi/d \tag{3.96}
$$

and μ, ν are non-negative integers such that

$$\ell_{-\mu-1} < -k < \ell_{-\mu}, \qquad \ell_\nu < k < \ell_{\nu+1}. \tag{3.97}$$

It can be then be shown, from (3.77)–(3.81), that

$$\varphi \sim \begin{cases} e^{i\alpha x + i\ell y} + \displaystyle\sum_{p=-\mu}^{\nu} R_p\, e^{i\ell_p y - ikxt_p} & \text{as } x \to -\infty, \\[2ex] \displaystyle\sum_{p=-\mu}^{\nu} T_p\, e^{i\ell_p y + ikxt_p} & \text{as } x \to +\infty, \end{cases} \tag{3.98}$$

where

$$R_p = (kdt_p)^{-1}(P_p + iQ_p), \qquad T_p = \delta_{0p} + (kdt_p)^{-1}(P_p - iQ_p) \tag{3.99}$$

and

$$P_p = \sum_{n=0}^{\infty} \left[a_{2n} c_{2n}(t_p) - i\,\mathrm{sgn}(\ell_p) a_{2n+1} c_{2n+1}(t_p) \right], \tag{3.100}$$

$$Q_p = \sum_{n=1}^{\infty} \left[b_{2n-1} s_{2n-1}(t_p) - i\,\mathrm{sgn}(\ell_p) b_{2n} s_{2n}(t_p) \right]. \tag{3.101}$$

The far-field solution is thus seen to consist of the incident plane wave plus a finite sum of reflected and transmitted plane waves with amplitudes $|R_p|$, $|T_p|$, $p = -\mu, \dots, \nu$. The transmitted waves are plane waves of the form $\exp(i\ell_p y + ikxt_p)$ and these make angles θ_p, where $\theta_p \in (-\pi/2, \pi/2)$, with the positive x-axis where, from equations (3.96) and (3.97), $\cos\theta_p = t_p$, $\sin\theta_p = \ell_p/k$. Thus using (3.73) we have

$$\sin\theta_p = \sin\beta + p\pi/kd, \qquad p = -\mu, \dots, \nu. \tag{3.102}$$

The reflected waves make angles $\pi - \theta_p$ with the positive x-axis.

Applying Green's theorem to $\varphi - \varphi_{\mathrm{I}}$ and its complex conjugate, see, for example, Achenbach et al. (1988), leads (exactly as in the derivation of equation 1.62) to a result which represents the conservation of energy:

$$\sum_{p=-\mu}^{\nu} t_p(|R_p|^2 + |T_p|^2) = t_0. \tag{3.103}$$

It is straightforward to obtain the behaviour of R_0 and T_0 in the limit as $kd \to 0$ with a/d fixed or as $a/d \to 0$ with kd fixed. From (3.93) and

(3.94), using (A.19)–(A.22), it can be shown that in either of the above limits

$$a_0 \sim -\tfrac{1}{4}\pi\mathrm{i}(ka)^2, \quad a_1 \sim -\tfrac{1}{2}\pi(ka)^2\sin\beta, \quad b_1 \sim -\tfrac{1}{2}\pi(ka)^2\cos\beta$$

and $a_n, b_n = O((ka)^4)$, $n \geq 2$. We thus obtain

$$R_0 \sim -\frac{\pi\mathrm{i}ka^2}{4d\cos\beta}(1 + 2\cos 2\beta), \quad T_0 \sim 1 + \frac{\pi\mathrm{i}ka^2}{4d\cos\beta}. \tag{3.104}$$

When just one mode is present (i.e. $\mu = \nu = 0$) these approximations can be improved by applying the conservation of energy condition (3.103) as described by Miles (1982).

Bibliographical notes

When $\ell = 0$ the multipoles used above are appropriate to the physical problem of a body on the centre-line of a water-wave channel. Similar multipoles can be derived for singularities which are off the centre-line in such a channel. This was done in McIver and Bennett (1993) and such multipoles were used to solve the scattering problem for an arbitrary finite array of circular cylinders in a channel in McIver and Linton (1994) and Linton and McIver (1996). Utsunomiya and Eatock Taylor (1999) used these multipoles to compute frequencies at which trapped modes exist in the vicinity of a row of circular cylinders placed across a channel. Fully three-dimensional problems concerning submerged spheres in channels have also been solved using multipoles (Wu 1998, Ursell 1999). An in-depth discussion of the use of multipoles in channel problems can be found in Linton (1997).

Multipoles equivalent to those used in this section but with $\ell > k$ were used in McIver, Linton, and McIver (1998) to find the possible wavenumbers at which pure Rayleigh-Bloch surface waves can propagate along an infinite row of circular cylinders. A Rayleigh-Bloch wave propagates along the surface of a structure that is adjacent to an infinite medium, with decay of the motion into the medium. For further details of such waves see §5.2.3.

Chapter 4

Integral equations

The formulation of wave scattering and radiation problems in terms of integral equations has proved to be of great utility. In a few cases such a formulation has resulted in an exact solution. However, integral equations may also be used to obtain asymptotic approximations and to form the basis of accurate numerical methods. In this chapter we concentrate on the derivation of integral equations from a boundary-value problem and on the numerical solution of such equations. The reader might consult Porter and Stirling (1990) for detailed discussion of the theory of integral equations and Delves and Mohamed (1985) for more on the numerical solution of integral equations.

The main problem to be solved here using integral equations is the radiation of waves by an obstacle with a prescribed distribution of velocity on its surface. This includes the case of wave scattering as the velocity distribution then arises from the incident wave. Two main methods of derivation are described. First of all, in §4.1 the solution is written as a distribution of wave sources over the surface of the obstacle and an integral equation is obtained for the unknown source strength. The second approach, described in §4.2, uses Green's theorem to obtain an integral equation. In the case of a thin obstacle a modification of the approach is required and this is given in §4.3. Specific applications to interior eigenvalue problems and free-surface problems are discussed in §§4.4 and 4.5, respectively.

One integral-equation approach to free-surface problems based on Green's theorem involves the use of rather complicated Green's functions. In applications where many different evaluations of a Green's function is required it is important to use an efficient algorithm. Algorithms for a variety of different problems are discussed in §4.6.

In §4.7, many of the ideas are illustrated in detail by consideration of the complementary problems of diffraction of water waves by a break-water with a gap and a breakwater of finite length. These particular problems are chosen as there is the interesting feature that 'embedding' formulas may be used to obtain the solutions to both problems for an arbitrary angle of wave incidence in terms of the gap problem for one angle of wave incidence. Two numerical methods are described for the solution of the gap problem and comparisons made between them. The relationship between these complementary problems is an example of Babinet's principle and this is discussed in §4.8.

4.1 Source distributions

To illustrate the main ideas behind the formulation of integral equations using source distributions, the following two-dimensional problem is considered. Let D denote the unbounded region exterior to the simple closed curve S_B; upper case letters P and Q are used to denote points in D and lower case letters p and q are used to denote points on the curve S_B. The problem is to determine $\phi(P)$ which satisfies the Helmholtz equation

$$(\nabla^2 + k^2)\phi(P) = 0, \quad P \in D, \tag{4.1}$$

the boundary condition

$$\frac{\partial \phi}{\partial n_p}(p) = f(p), \quad p \in S_B, \tag{4.2}$$

and the radiation condition

$$r_P^{1/2}\left(\frac{\partial \phi}{\partial r_P}(P) - \mathrm{i}k\phi(P)\right) \to 0 \quad \text{as} \quad r_P \to \infty. \tag{4.3}$$

Here $f(p)$ is a given function on S_B, k is a given positive real number, and n_p is a normal coordinate at $p \in S_B$ measured in the direction from D towards S_B. The origin of coordinates is within D_-, the interior of S_B, and (r_P, θ_P) are the polar coordinates of P relative to O. The function $f(p)$ may be an imposed velocity distribution or may arise from an incident wave. Thus the formulation includes both scattering and radiation problems. A boundary condition of the form (4.2) is referred to as a Neumann boundary condition. Note that, unless otherwise stated, references in this text to a Neumann boundary condition are to the homogeneous case for which $f(p) \equiv 0$.

The fundamental source solution for the Helmholtz equation is

$$G(P, Q) = \frac{1}{4\mathrm{i}} H_0^{(1)}(k R_{PQ}), \quad P \neq Q, \tag{4.4}$$

where $H_0^{(1)}$ denotes a Hankel function and R_{PQ} is the distance of the field point P from the source point Q. Note from (A.17) that

$$G(P, Q) \sim \frac{1}{2\pi} \ln R_{PQ} \quad \text{as} \quad k R_{PQ} \to 0. \tag{4.5}$$

Now

$$
\begin{aligned}
R_{PQ} &= \left[r_P^2 + r_Q^2 - 2 r_P r_Q \cos(\theta_P - \theta_Q) \right]^{1/2} \\
&= r_P \left[1 - \frac{r_Q}{r_P} \cos(\theta_P - \theta_Q) + O\left(\frac{r_Q^2}{r_P^2} \right) \right] \quad \text{as} \quad \frac{r_P}{r_Q} \to \infty \tag{4.6}
\end{aligned}
$$

and hence from equation (A.12) as $k r_P \to \infty$,

$$H_0^{(1)}(k R_{PQ}) \sim \left(\frac{2}{\pi k r_P} \right)^{1/2} \mathrm{e}^{\mathrm{i}(k r_P - k r_Q \cos(\theta_P - \theta_Q) - \pi/4)} \tag{4.7}$$

so that $G(P, Q)$ also satisfies the radiation condition (4.3). A solution of (4.1)–(4.3) is sought in the form

$$\phi(P) = \int_{S_\mathrm{B}} \mu(q) G(P, q) \, \mathrm{d}s_q, \tag{4.8}$$

which by the properties of $G(P, Q)$ satisfies both (4.1) and (4.3). It remains to satisfy (4.2) by choice of the 'source distribution' $\mu(q)$. Consider differentiation of $\phi(P)$ with respect to the normal direction at $p \in S_\mathrm{B}$. For $P \notin S_\mathrm{B}$

$$\frac{\partial \phi}{\partial n_p}(P) = \int_{S_\mathrm{B}} \mu(q) \frac{\partial G}{\partial n_p}(P, q) \, \mathrm{d}s_q, \tag{4.9}$$

but for $P \to p \in S_\mathrm{B}$,

$$\frac{\partial \phi}{\partial n_p}(p) = -\tfrac{1}{2}\mu(p) + \int_{S_\mathrm{B}} \mu(q) \frac{\partial G}{\partial n_p}(p, q) \, \mathrm{d}s_q. \tag{4.10}$$

(see Smirnov 1964, §195 and §199). Equation (4.8) may be used to continue ϕ in to the interior D_- of S_B. The potential is continuous across S_B, but its normal derivative has a discontinuity and in particular

$$\frac{\partial \phi}{\partial n_p}(p_-) = \tfrac{1}{2}\mu(p) + \int_{S_B} \mu(q)\frac{\partial G}{\partial n_p}(p,q)\,ds_q, \qquad (4.11)$$

where $p_- \in S_B$ is obtained by approaching S_B from D_-.

By construction (4.8) satisfies the boundary condition (4.2) provided the source distribution $\mu(q)$ satisfies the integral equation

$$-\tfrac{1}{2}\mu(p) + \int_{S_B} \mu(q)\frac{\partial G}{\partial n_p}(p,q)\,ds_q = f(p), \quad p \in S_B. \qquad (4.12)$$

This is a Fredholm integral equation of the second kind for the source distribution $\mu(q)$. Once $\mu(q)$ is determined the potential everywhere in D follows from (4.8).

It might seem at first sight that the kernel of (4.12) is singular. However, Smirnov (1964, §199) demonstrates that for logarithmic G, then the kernel of an integral equation of the form (4.12) is a continuous function of the arc length s (see also §4.1.1 below). This is not true in three-dimensional problems for which the kernel is weakly singular (see Kellogg 1953, p. 299).

4.1.1 Numerical solution

One approach to the numerical solution of (4.12) is now described. Suppose that the curve S_B may be described in terms of a parameter $u \in [a,b]$ so that (4.12) may be rewritten as

$$-\tfrac{1}{2}M(u) + \int_a^b M(v)\frac{\partial \mathcal{G}}{\partial n_p}(u,v)\,w(v)\,dv = f(u), \quad u \in [a,b], \qquad (4.13)$$

where $(x(u), y(u))$ are the coordinates of the field point p, $(x(v), y(v))$ are the coordinates of the field point q,

$$M(u) \equiv \mu(p), \quad \mathcal{G}(u,v) \equiv G(p,q), \quad \frac{\partial \mathcal{G}}{\partial n_p}(u,v) \equiv \frac{\partial G}{\partial n_p}(p,q) \qquad (4.14)$$

and

$$w(v) = \left([x'(v)]^2 + [y'(v)]^2\right)^{1/2}. \qquad (4.15)$$

The interval $[a, b]$ is divided into N equal subintervals with end points $\{u_i = a + i(b-a)/N;\ i = 0, 1, \dots, N\}$ and midpoints $\{\hat{u}_i = (u_{i-1} + u_i)/2;\ i = 1, 2, \dots, N\}$. Collocation at the element midpoints then leads to the simultaneous equations

$$-\tfrac{1}{2}M(\hat{u}_i) + \frac{b-a}{N} \sum_{j=1}^{N} M(\hat{u}_j) \frac{\partial \mathcal{G}}{\partial n_p}(\hat{u}_i, \hat{u}_j)\, w(\hat{u}_j)$$

$$= f(\hat{u}_i), \quad i = 1, 2, \dots, N, \quad (4.16)$$

for the source distribution values $\{M(\hat{u}_i);\ i = 1, 2, \dots, N\}$.

As noted after (4.12), the normal derivative of the Green's function is non-singular when the source point q and field point p coincide and the limiting value as $q \to p$ may be found as follows. Write

$$G(p, q) = \frac{1}{2\pi} \ln R_{pq} + \tilde{G}(p, q), \qquad (4.17)$$

where $\tilde{G}(p, q)$ and its first derivatives are non-singular when $q = p$, and hence its normal derivative may be calculated without difficulty everywhere on S_{B}. In terms of the parameters u and v

$$\frac{\partial}{\partial n_p}(\ln R_{pq}) = \frac{-y'(u)[x(u) - x(v)] + x'(u)[y(u) - y(v)]}{w(u)R_{pq}^2}. \qquad (4.18)$$

By Taylor's theorem

$$x(v) = x(u) + (v-u)x'(u) + \tfrac{1}{2}(v-u)^2 x''(u_1),$$
$$y(v) = y(u) + (v-u)y'(u) + \tfrac{1}{2}(v-u)^2 y''(u_2), \qquad (4.19)$$

where u_1 and u_2 lie between u and v, and hence as $|u - v| \to 0$

$$R_{pq} = \left[(x(u) - x(v))^2 + (y(u) - y(v))^2\right]^{1/2}$$
$$= |v - u|w(u) + O(|v - u|^2) \qquad (4.20)$$

and so as $kR_{pq} \to 0$

$$\frac{\partial}{\partial n_p}(\ln R_{pq}) \to \frac{y'(u)x''(u) - x'(u)y''(u)}{2[w(u)]^3}. \qquad (4.21)$$

Once the source distribution is determined (4.8) may be used to determine the field throughout D. Thus

$$\phi(P) = \frac{b-a}{N} \sum_{j=1}^{N} M(\hat{u}_j) \int_{u_{j-1}}^{u_j} G(P, q(v))w(v)\, dv, \qquad (4.22)$$

where the integral of G has been retained because of the logarithmic singularity as $P \to q$ if P is chosen on S_B.

4.1.2   Irregular values

The boundary-value problem (4.1)–(4.3) is known to have a unique solution for all $k^2 > 0$ (Colton and Kress 1983, §3.3). A difficulty with the above method is that the integral equation (4.12) is not uniquely solvable for certain values of k^2 known as irregular values (a phenomenon recognized by Lamb 1932, §290). These values of k^2 correspond to eigenvalues of the interior Dirichlet problem specified by

$$(\nabla^2 + k^2)\phi(P_-) = 0, \quad P_- \in D_-, \tag{4.23}$$

$$\phi(p_-) = 0, \quad p_- \in S_B, \tag{4.24}$$

where D_- is the interior of S_B and p_- is obtained by approaching S_B from D_-. A condition of the form (4.24) is referred to as a (homogeneous) Dirichlet boundary condition.

The correspondence between the irregular values and the eigenvalues of an interior problem can be seen from an examination of the homogeneous equation corresponding to (4.12). From the Fredholm alternative, if the homogeneous equation has no non-trivial solutions then the non-homogeneous equation has a unique solution (see, for example, Porter and Stirling 1990, theorem 4.10).

Suppose that the homogeneous equation

$$-\tfrac{1}{2}\mu(p) + \int_{S_B} \mu(q) \frac{\partial}{\partial n_p} G(p, q) \, ds_q = 0, \quad p \in S_B \tag{4.25}$$

does have a non-trivial solution for $\mu(p)$. The corresponding potential defined by (4.8) has zero normal derivative on S_B and hence, by the uniqueness of the solution to the boundary-value problem, $\phi(P) = 0$ for all $P \in D$ and in particular $\phi(p) = 0$ for $p \in S_B$. Further, by continuity of the representation (4.8), $\phi(p_-) = \phi(p) = 0$ for $p_- \in S_B$ (see equation 4.10) and so, as long as k^2 is not an eigenvalue of the interior Dirichlet problem, $\phi(P_-) = 0$ for all $P_- \in D_-$ and, in particular, from (4.11)

$$\frac{\partial \phi}{\partial n_p}(p_-) = \tfrac{1}{2}\mu(p) + \int_{S_B} \mu(q) \frac{\partial}{\partial n_p} G(p, q) \, ds_q = 0. \tag{4.26}$$

Hence, from (4.25), $\mu(p) = 0$ contradicting the original assumption that $\mu(p) \neq 0$. Thus, if k^2 is not an eigenvalue of the interior Dirichlet problem (4.23)–(4.24), then the integral equation (4.12) is uniquely solvable.

Conversely, if k^2 is an eigenvalue of the interior Dirichlet problem and $\phi(P_-)$ is a corresponding non-trivial solution, then from (4.11) and (4.25),

$$\mu(p_-) = \frac{\partial \phi}{\partial n_p}(p_-) \neq 0 \qquad (4.27)$$

and there is no contradiction in assuming a non-trivial solution for μ.

Thus, a solution of the original boundary-value problem based on the integral equation (4.12) may break down if k^2 is an eigenvalue of the Laplacian for the interior Dirichlet problem. This is purely an artifact of the solution procedure and, more specifically, of the Green's function used. One way of eliminating irregular values (Ursell 1973) is to modify the Green's function by adding terms that are singular at an origin O within S_{B}. Let $S_{\mathrm{B}}{}'$ be a contour within S_{B} and surrounding O. The additional terms in the Green's function are constructed to ensure that the interior potential $\phi(P_-)$ satisfies a condition on $S_{\mathrm{B}}{}'$ which ensures that if $\phi(p_-) = 0$, $p_- \in S_{\mathrm{B}}$, then $\phi(P_-) \equiv 0$ when P_- lies between S_{B} and $S_{\mathrm{B}}{}'$. This eliminates the possibility of non-trivial solutions to an interior Dirichlet problem and hence the modified integral equation is uniquely solvable. An alternative is to use a method that is known to be free of irregular values such as the null-field method (Martin 1980).

4.2 Green's theorem

Consider a region D bounded by a surface S and suppose that throughout D the complex-valued functions ϕ and G are continuously differentiable and have continuous second partial derivatives. It follows that ϕ and G satisfy Green's second identity (Kellogg 1953, p. 215),

$$\iiint\limits_{D} (\phi \nabla^2 G - G \nabla^2 \phi) \, \mathrm{d}V = \iint\limits_{S} \left(\phi \frac{\partial G}{\partial n} - G \frac{\partial \phi}{\partial n} \right) \mathrm{d}S, \qquad (4.28)$$

where n is a coordinate directed in the outward normal direction to the surface S. Further, if ϕ and G are both solutions of the Helmholtz equation

$$\nabla^2 U + k^2 U = 0 \tag{4.29}$$

throughout D for some k (possibly zero), then the integrand within the volume integral vanishes identically and (4.28) reduces to

$$\iint\limits_{S} \left(\phi \frac{\partial G}{\partial n} - G \frac{\partial \phi}{\partial n} \right) dS = 0. \tag{4.30}$$

This relation is often referred to as Green's theorem.

First of all suppose that $\phi(P)$, $P \in D$, is the unknown function in a particular two-dimensional problem so that S is a closed curve (following the notation of §4.1, we use P and Q to denote points within D, and p and q to denote points on the boundary of D). By appropriate choice of a Green's function $G(P, Q)$, (4.30) may be used to obtain an integral representation for ϕ, and/or an integral equation for ϕ, in terms of the behaviour of ϕ on S. The appropriate choice for G has the asymptotic form

$$G(P, Q) \sim \frac{1}{2\pi} \ln R_{PQ} \quad \text{as} \quad R_{PQ} \to 0, \tag{4.31}$$

where R_{PQ} is the distance of the field point P from the source point Q. By construction $G(P, Q)$ is singular at Q and so, for the application of Green's theorem, if Q lies within or on S then it must be excluded from the domain D. Consider first of all the case when Q is within S and let S_ϵ be a circle of radius ϵ with centre at Q. Application of Green's theorem is now made to ϕ and G over the domain between S and S_ϵ so that the line integration is over $S \cup S_\epsilon$. In the limit as $\epsilon \to 0$, the contribution from S_ϵ is

$$\lim_{\epsilon \to 0} \int_{S_\epsilon} \left(\phi(p) \frac{\partial G}{\partial n_p}(p, Q) - G(p, Q) \frac{\partial \phi}{\partial n_p}(p) \right) ds_p$$

$$= \lim_{\epsilon \to 0} \int_0^{2\pi} \left(\phi(p) \left[-\frac{\partial G}{\partial R_{pQ}}(p, Q) \right] \right.$$

$$\left. - G(p, Q) \left[-\frac{\partial \phi}{\partial R_{pQ}}(p) \right] \right) \Bigg|_{R_{pQ} = \epsilon} \epsilon \, d\theta$$

$$= \lim_{\epsilon \to 0} \int_0^{2\pi} \left(\phi(p) \left[-\frac{1}{2\pi R_{pQ}} \right] - \frac{\ln R_{pQ}}{2\pi} \left[-\frac{\partial \phi}{\partial R_{pQ}}(p) \right] \right) \Bigg|_{R_{pQ} = \epsilon} \epsilon \, d\theta$$

$$= -\phi(Q), \tag{4.32}$$

where the asymptotic form (4.31) and $\lim_{\epsilon \to 0} \phi(p) = \phi(Q)$ have been used in the limiting process. Hence, Green's theorem (4.30) yields

$$\phi(Q) = \int_S \left(\phi(p) \frac{\partial G}{\partial n_p}(p, Q) - G(p, Q) \frac{\partial \phi}{\partial n_p}(p) \right) ds_p, \quad Q \in D, \quad (4.33)$$

which is an integral representation for ϕ at an arbitrary point within D in terms of the values of ϕ and its normal derivative on the boundary. The subscripts on the normal coordinate n_p and the line element ds_p indicate that these are associated with the point $p \in S$. If ϕ is known on S, and G can be constructed so that $G(p, Q) = 0$ for all $p \in S$, then (4.33) gives the solution explicitly. Alternatively, if $\partial \phi(p)/\partial n_p$ is known for $p \in S$ and G can be constructed so that $\partial G(p, Q)/\partial n_p = 0$ for all $p \in S$, then again (4.33) gives the solution.

If q lies on the boundary of S then a semicircular (rather than a circular) contour is required to exclude q from D which results in

$$\tfrac{1}{2}\phi(q) = \int_S \left(\phi(p) \frac{\partial G}{\partial n_p}(p, q) - G(p, q) \frac{\partial \phi}{\partial n_p}(p) \right) ds_p, \quad q \in S. \quad (4.34)$$

If for $p \in S$, $\partial \phi(p)/\partial n_p$ is known, then (4.34) is an integral equation for the boundary values of ϕ. As noted in §4.1, the normal derivative of G appearing in (4.34) is a continuous function of the arc length when G has a logarithmic singularity.

Simplifications are often obtained by constructing G to satisfy the same boundary condition as ϕ on some $S_0 \subset S$ so that the contribution to the integral from S_0 vanishes. Once the integral equation has been solved for ϕ, (4.33) may be used to determine ϕ throughout D. It should also be noted that if the source point Q lies outside D then

$$\int_S \left(\phi(p) \frac{\partial G}{\partial n_p}(p, Q) - G(p, Q) \frac{\partial \phi}{\partial n_p}(p) \right) ds_p = 0. \quad (4.35)$$

In the above derivation, the singular point of the Green's function was regarded as fixed. A change of notation and an application of the symmetry property $G(P, Q) = G(Q, P)$, which holds for the particular problems under consideration here, yields

$$\phi(P) = \int_S \left(\phi(q) \frac{\partial G}{\partial n_q}(P, q) - G(P, q) \frac{\partial \phi}{\partial n_q}(q) \right) ds_q, \quad P \in D, \quad (4.36)$$

and

$$\tfrac{1}{2}\phi(p) = \int_S \left(\phi(q) \frac{\partial G}{\partial n_q}(p, q) - G(p, q) \frac{\partial \phi}{\partial n_q}(q) \right) ds_q, \quad p \in S. \quad (4.37)$$

The representations (4.36)–(4.37) may now be interpreted in terms of distributions of sources and dipoles over the boundary surface S.

In three dimensions similar formulas may be obtained. The appropriate Green's function has the asymptotic form

$$G(P, Q) \sim -\frac{1}{4\pi R_{PQ}} \quad \text{as} \quad R_{PQ} \to 0. \tag{4.38}$$

When P lies inside (or on) S it is excluded from D by a sphere (or hemisphere) and this results in equations identical in form to those for the two-dimensional case in equations (4.36)–(4.37).

4.2.1   Scattering by a vertical cylinder

To illustrate the integral-equation formulation described in §4.2, the problem of the scattering of waves by a vertical cylinder extending throughout water of constant depth is considered. The cylinder has a uniform but otherwise arbitrary cross-section which is denoted by S_B and the corresponding cross-section of the fluid region is denoted by D. The origin of coordinates O is chosen to be within S_B and (r_P, θ_P) are horizontal polar coordinates of a point P relative to O. The case in which the cross-section is a circle was solved using an eigenfunction expansion in §2.4.1 and as in that case, the depth may be removed from the problem by writing the potential as

$$\Phi(x, y, z, t) = \mathrm{Re}\left\{ [\phi_\mathrm{I}(x) + \phi(x, y)]\, \psi_0(z)\, \mathrm{e}^{-\mathrm{i}\omega t} \right\}, \tag{4.39}$$

where the incident wave potential is

$$\phi_\mathrm{I} = \mathrm{e}^{\mathrm{i}kx}. \tag{4.40}$$

The diffracted wave potential ϕ satisfies the Helmholtz equation (4.1), the cylinder boundary condition

$$\frac{\partial \phi}{\partial n_p}(p) = -\frac{\partial \phi_\mathrm{I}}{\partial n_p}, \quad p \in S_\mathrm{B}, \tag{4.41}$$

where n_p is the normal directed out of the fluid region at p, and the radiation condition (4.3). The problem above is a special case of that used for illustrative purposes in §4.1.

The fundamental Green's function for the Helmholtz equation is given by equation (4.4). The contour S for the application of Green's theorem is chosen to be $S_\mathrm{B} \cup S_X$ where S_X is a circle of large radius $r_P = X$.

Now as $kr_P \to \infty$ for fixed Q the asymptotic form of $G(P, Q)$ follows from (4.7) and is

$$G(P, Q) \sim \hat{G}(P, Q) r_P^{-1/2} \, e^{ikr_P}, \qquad (4.42)$$

where $\hat{G}(P, Q)$ is independent of r_P. In the limit $kX \to \infty$, the contribution to the integral on the right-hand side of (4.33) is therefore

$$\lim_{X \to \infty} \int_0^{2\pi} \left(\phi(p) \frac{\partial G}{\partial r_p}(p, Q) - G(p, Q) \frac{\partial \phi}{\partial r_p}(p) \right) \bigg|_{r_p = X} X \, d\theta$$

$$= \lim_{X \to \infty} \int_0^{2\pi} \left(ik\phi(p) - \frac{\partial \phi}{\partial r_p}(p) \right) \bigg|_{r_p = X} \hat{G}(p, Q) \, e^{ikX} X^{1/2} d\theta \quad (4.43)$$

which vanishes by virtue of the radiation condition (4.3). The integral equation (4.37) therefore reduces to

$$-\tfrac{1}{2}\phi(p) + \int_{S_{\rm B}} \phi(q) \frac{\partial G}{\partial n_q}(p, q) \, ds_q$$

$$= -\int_{S_{\rm B}} G(p, q) \frac{\partial \phi_{\rm I}}{\partial n_q}(q) \, ds_q, \quad p \in S_{\rm B}, \quad (4.44)$$

which is a Fredholm integral equation of the second kind for the values of ϕ on the cylinder contour $S_{\rm B}$.

A simplification of (4.44) may be obtained as follows. An application of Green's theorem (4.30) in D_-, the domain interior to $S_{\rm B}$, to the incident wave $\phi_{\rm I} = \exp(ikx)$ and the Green's function $G(P, Q)$ yields

$$\tfrac{1}{2}\phi_{\rm I}(p) = -\int_{S_{\rm B}} \left(\phi_{\rm I}(p) \frac{\partial G}{\partial n_q}(p, q) - G(p, q) \frac{\partial \phi_{\rm I}}{\partial n_q}(q) \right) ds_q, \quad p \in S_{\rm B},$$

$$(4.45)$$

where the normal coordinate n_q is directed into D_-. Elimination between (4.44) and (4.45) of the term involving the derivative of $\phi_{\rm I}$ yields

$$-\tfrac{1}{2}\psi(p) + \int_{S_{\rm B}} \psi(q) \frac{\partial G}{\partial n_q}(p, q) \, ds_q = -\phi_{\rm I}(p), \quad p \in S_{\rm B}, \quad (4.46)$$

where $\psi = \phi + \phi_{\rm I}$. A numerical scheme based on this formulation is given by Isaacson (1978) and he includes an investigation of the interactions between neighbouring cylinders.

The kernel of the integral equations (4.44) and (4.46) is the transpose of that appearing in (4.12) and so, by the Fredholm alternative theorem, it will therefore also suffer from the problem of irregular values as described in §4.1.2.

4.3   Thin obstacles

If the obstacle under consideration has negligible thickness, then the integral equation technique described above requires modification. Consider the two-dimensional problem (4.1)–(4.3) used for illustration in §4.1, but where the obstacle is now taken to be thin. Denote the obstacle by the arc Γ, the two sides of Γ by $\Gamma^{\pm}$, and the corresponding normal coordinates directed out of the fluid by $n_p^{\pm}$ for $p \in \Gamma$. The boundary condition (4.2) is rewritten as

$$\frac{\partial \phi}{\partial n_p^{\pm}}(p) = \pm f(p), \quad p \in \Gamma \tag{4.47}$$

(this class of forcing functions includes the case of an oscillating rigid plate). An application of Green's theorem to ϕ and the fundamental source (4.4) and use of the radiation condition leads to a representation for ϕ in the form (4.36) where the integration is over both sides of the arc Γ. Now by definition

$$\frac{\partial}{\partial n_q^-} = -\frac{\partial}{\partial n_q^+}, \quad q \in \Gamma, \tag{4.48}$$

and hence after application of (4.47) the representation (4.36) reduces to

$$\phi(P) = \int_{\Gamma} [\phi(q)] \frac{\partial G}{\partial n_q^+}(P, q) \, ds_q, \quad P \in D, \tag{4.49}$$

where

$$[\phi(q)] = \phi(q^+) - \phi(q^-), \quad q^{\pm} \in \Gamma^{\pm}, \tag{4.50}$$

is the jump in the potential across Γ and some cancellation has occurred in the integrations along $\Gamma^{\pm}$. It remains to satisfy the boundary condition (4.47). Now the two equations

$$\frac{\partial \phi}{\partial n_p^{\pm}}(p) = \frac{\partial}{\partial n_p^{\pm}} \int_{\Gamma} [\phi(q)] \frac{\partial G}{\partial n_q^+}(p, q) \, ds_q = \pm f(p), \quad p \in \Gamma, \tag{4.51}$$

both reduce to the same equation by virtue of (4.48) and hence the superscripts may be dropped to obtain

$$\frac{\partial}{\partial n_p} \int_{\Gamma} [\phi(q)] \frac{\partial G}{\partial n_q}(p, q) \, ds_q = f(p), \quad p \in \Gamma, \tag{4.52}$$

which is an integral equation for $[\phi(q)]$. For definiteness all normal derivatives will be associated with Γ^+ from now on. If $[\phi(q)]$ is determined for all $q \in \Gamma$ then (4.49) yields the potential ϕ throughout the fluid domain.

By the theorem in §2 of Martin and Rizzo (1989), (4.52) may be rewritten as

$$\oint_\Gamma [\phi(q)] \frac{\partial^2 G}{\partial n_p \partial n_q}(p, q) \, ds_q = f(p), \quad p \in \Gamma, \tag{4.53}$$

where the integral is now a finite-part integral as defined in Appendix C. An equation of this type is known as a hypersingular integral equation because of the strength of the kernel singularity. Denote the unit normals at $p \in \Gamma$ by $\mathbf{n}(p) = (n_1^p, n_2^p)$. If $p \in \Gamma$ has coordinates (x, y) and $q \in \Gamma$ has coordinates (ξ, η), then

$$\frac{\partial^2 G}{\partial n_p \partial n_q} = \left(n_1^p \frac{\partial}{\partial x} + n_2^p \frac{\partial}{\partial y} \right) \left(n_1^q \frac{\partial}{\partial \xi} + n_2^q \frac{\partial}{\partial \eta} \right) G$$

$$= \frac{k \, \mathbf{n}(p) \cdot \mathbf{n}(q)}{4\mathrm{i} R_{pq}} H_1^{(1)}(k R_{pq})$$

$$- \frac{k^2 (\mathbf{n}(p) \cdot \mathbf{R}_{pq})(\mathbf{n}(q) \cdot \mathbf{R}_{pq})}{4\mathrm{i} R_{pq}^2} H_2^{(1)}(k R_{pq}). \tag{4.54}$$

where $\mathbf{R}_{pq}$ is the position vector of p relative to q, $R_{pq} = |\mathbf{R}_{pq}|$, and $H_m^{(1)}$ denotes the Hankel function of the first kind with order m.

4.3.1 Numerical solution

To illustrate a numerical procedure for solving the equation (4.53) consider the case when Γ is a flat plate of length $2a$ inclined at an angle α to the x axis and with its centre at the origin. Parameterize Γ by taking

$$\Gamma = \{(ua \cos \alpha, ua \sin \alpha), \ u \in [-1, 1]\}, \tag{4.55}$$

so that

$$\mathbf{n}(p) = \mathbf{n}(q) = (-\sin \alpha, \cos \alpha) \tag{4.56}$$

and in particular

$$\mathbf{n}(p) \cdot \mathbf{R}_{pq} = \mathbf{n}(q) \cdot \mathbf{R}_{pq} = 0. \tag{4.57}$$

The integral equation (4.53) then reduces to

$$\frac{k}{4\mathrm{i}} \fint_{-1}^{1} P(v) \frac{H_1^{(1)}(ka|u-v|)}{|u-v|} \, \mathrm{d}v = F(u), \quad u \in (-1,1), \qquad (4.58)$$

where $F(u) \equiv f(p)$, $P(v) \equiv [\phi(q)]$ and $P(\pm 1) = 0$. Now from (A.18) it is appropriate to write

$$\frac{H_1^{(1)}(ka|u-v|)}{|u-v|} = \frac{2}{\pi i k a} \left\{ \frac{1}{(u-v)^2} + L(u,v) \right\}, \qquad (4.59)$$

where $L(u,v)$ has only a logarithmic singularity as $|u-v| \to 0$ (see Abramowitz and Stegun 1965, equation 9.1.88), and hence the integral equation is rewritten as

$$\frac{1}{2\pi a} \fint_{-1}^{1} P(v) \left\{ \frac{1}{(u-v)^2} + L(u,v) \right\} \mathrm{d}v = -F(u), \quad u \in (-1,1). \quad (4.60)$$

The integral operator $\mathcal{K}$ defined by

$$(\mathcal{K}g)(u) \equiv \fint_{-1}^{1} \frac{(1-v^2)^{1/2} g(v)}{(u-v)^2} \, \mathrm{d}v \qquad (4.61)$$

has eigenfunctions which are the second-kind Chebyshev polynomials $U_n(v)$, $n = 0, 1, 2, \ldots$. In particular

$$\fint_{-1}^{1} \frac{(1-v^2)^{1/2} U_n(v)}{(u-v)^2} \, \mathrm{d}v = -\pi(n+1) U_n(u) \qquad (4.62)$$

(Frenkel 1983, equation 33). The Chebyshev polynomials form a complete set over the interval $[-1,1]$ and hence it is appropriate to look for an approximate solution of (4.60) in the form

$$P(v) \approx (1-v^2)^{1/2} \sum_{n=0}^{N} a_n U_n(v), \qquad (4.63)$$

where $\{a_n; \; n = 0, 1, \ldots, N\}$ are coefficients to be determined; this form reflects the known asymptotic behaviour of $P(v)$ as the end points of $(-1,1)$ are approached (Martin 1991a). Substitution of the approximation (4.63) into the integral equation (4.60) yields

$$\sum_{n=0}^{N} a_n A_n(u) = -2\pi a F(u), \quad u \in (-1,1), \qquad (4.64)$$

where

$$A_n(u) = -\pi(n+1)U_n(u) + \int_{-1}^{1}(1-v^2)^{1/2}U_n(v)L(u,v)\,dv \qquad (4.65)$$

and (4.62) has been used. The integral involving $L(u,v)$ can be evaluated numerically without difficulty after suitable treatment of the logarithmic singularity. Parsons and Martin (1992) discuss a number of alternatives for the solution of (4.64) and choose to use a collocation scheme based on the zeros of the first-kind Chebyshev polynomials given by

$$u_j = \cos\frac{(2j+1)\pi}{2N+2}, \quad j = 0,1,\dots,N, \qquad (4.66)$$

so that (4.64) reduces to

$$\sum_{n=0}^{N} a_n A_n(u_j) = -2\pi a F(u_j), \quad j = 0,1,\dots,N. \qquad (4.67)$$

Once (4.67) has been solved for the coefficients $\{a_n;\ n = 0,1,\dots,N\}$ then (4.49) may be used to determine the wave field anywhere. For the case of the flat plate discussed above

$$\phi(x,y) = \int_{-1}^{1} P(v)\frac{\partial G}{\partial n_q}(x,y;va\cos\alpha,va\sin\alpha)a\,dv$$

$$\approx \sum_{n=0}^{N} a_n \int_{-1}^{1}(1-v^2)^{1/2}U_n(v)\frac{\partial G}{\partial n_q}(x,y;va\cos\alpha,va\sin\alpha)a\,dv$$

$$(4.68)$$

where

$$\frac{\partial G}{\partial n_q}(x,y;\xi,\eta) = \frac{k\,\mathbf{n}(q)\cdot\mathbf{R}_{Pq}}{4iR_{Pq}}H_1^{(1)}(kR_{Pq}) \qquad (4.69)$$

and $\mathbf{R}_{Pq}$ is the position vector of the field point (x,y) relative to the source point (ξ,η).

Bibliographical notes

The theoretical basis for hypersingular integral equations is discussed by Martin and Rizzo (1989, 1996) and the end-point behaviour of solutions to such equations by Martin (1991a). The expansion-collocation method described above for one-dimensional hypersingular equations is numerically very effective and its convergence in various function spaces has been proved by Golberg (1983, 1985) and Ervin and Stephan (1992).

Numerical results for an integral equation in the form (4.53), with the Green's function given by (4.4), are given by Frenkel (1983), although a slightly different approach to the numerical solution is adopted. The problem where the single barrier in the example described above is replaced by an infinite row of identical barriers was formulated as a hypersingular integral equation and solved, again using a modification to the expansion-collocation method, by Achenbach and Li (1986). Alternative numerical approaches to hypersingular integral equations are also described in Krishnasamy et al. (1990) and Guiggiani et al. (1992).

Various applications of hypersingular integral equations are briefly discussed by Martin (1991a). Applications to problems in water waves are given by Parsons and Martin (1992, 1994, 1995). In the first of these papers the scattering by submerged flat plates was considered and in the second the method was extended to include submerged curved plates and surface-piercing plates. The third paper concerned the trapping of waves above submerged plates. Linton and Kuznetsov (1997) used Parsons and Martin's method to numerically compute the frequencies of trapped modes between two surface-piercing angled barriers for the two-dimensional water-wave problem.

The hypersingular integral equation approach can also be used to solve three-dimensional problems involving thin surfaces. Thus Martin and Farina (1997) and Farina and Martin (1998) have derived two-dimensional hypersingular integral equations for radiation and scattering problems involving a submerged horizontal disk.

4.4   Interior problems

The integral-equation techniques described in sections §§4.1–4.2 are readily applied to interior problems. For example, one such problem is the generalization of the problem in §2.3.5 to sloshing in a cylindrical container of arbitrary cross-section S_B surrounding the two-dimensional region D. Let n_p be a normal coordinate to S_B at p directed out of D. The aim is to determine the eigenvalues k^2 of the Laplacian such that the problem

$$(\nabla^2 + k^2)\phi(P) = 0, \quad P \in D, \tag{4.70}$$

$$\frac{\partial \phi}{\partial n_p}(p) = 0, \quad p \in S_B, \tag{4.71}$$

has non-trivial solutions for ϕ. The integral equations obtained from the two approaches of §§4.1–4.2 are essentially the same; the only difference between them is that the kernel resulting from the source-distribution approach is the transpose of that obtained from the Green's theorem approach. The source-distribution approach leads to an equation in the form (4.25).

In solving such interior problems care has to be taken in the choice of Green's function. As the above eigenvalue problem is real valued (in particular, it does not involve a radiation condition) it is possible to formulate an integral equation using only the (singular) imaginary part

$$G_i(P, Q) = \frac{1}{4\pi} Y_0(k R_{PQ}) \tag{4.72}$$

of the Helmholtz source (4.4), where Y_0 denotes a Bessel function of the second kind. However, Mattioli (1980) has shown that a real Green's function leads to an integral equation that has irregular values corresponding to eigenvalues of an exterior Dirichlet problem, but these can be eliminated completely by using the complex form (4.4).

This can be explained as follows. The relevance of the exterior Dirichlet problem arises in a similar fashion to the interior Dirichlet problem described in §4.1.2. Suppose that G_i is used in the formulation of the integral equation for the source distribution $\mu(p)$, $p \in S_B$, and assume there is a non-trivial solution $\mu(p)$ of the homogeneous equation. Such a solution corresponds to a potential ϕ that is zero on S_B, but has nonzero normal derivative on the exterior of S_B. The integral representation for ϕ shows that it corresponds to a particular form of standing wave at infinity in the exterior domain. A given standing wave is consistent with a Dirichlet condition on S_B, and hence a contradiction is not obtained, only for certain values of k^2, the irregular values. However, if the complex G is used, then ϕ corresponds to outgoing waves at infinity that cannot be generated by a Dirichlet condition on S_B for any k^2, and therefore there are no irregular values.

4.5   Free-surface problems

Consider a problem in water waves involving the scattering of an incident wave by a fixed structure or the radiation of waves by an oscillating
structure in water of constant depth h. Let D denote the fluid domain,
S_B the surface of the obstacle, S_h the bed, and F the free surface. From
the theory given in §§1.2–1.3, the boundary-value problem requires the
velocity potential ϕ to satisfy Laplace's equation

$$\nabla^2 \phi(P) = 0, \quad P \in D, \tag{4.73}$$

the boundary conditions

$$\frac{\partial \phi}{\partial n_p}(p) = f(p), \quad p \in S_B, \tag{4.74}$$

$$\frac{\partial \phi}{\partial n_p}(p) = 0, \quad p \in S_h, \tag{4.75}$$

$$\frac{\partial \phi}{\partial n_p}(p) = K\phi(p), \quad p \in F, \tag{4.76}$$

and a radiation condition specifying outgoing waves (in the form of equation 1.28 or 1.29). Here, $f(p)$ represents the forcing from either the
incident wave or oscillating body.

The main additional difficulty in such problems is the reduction to an
integral equation over a finite boundary by a suitable choice of Green's
function. The usual procedure is to construct a Green's function $G(P,Q)$
that satisfies Laplace's equation for $P \neq Q$, the boundary conditions on
F and S_h, and the appropriate radiation condition. Suitable Green's
functions for a variety of situations are easily obtained from the results
for source potentials in Appendix B; care must be taken to ensure that
the strength of the singularity is adjusted in accordance with (4.31) or
(4.38). In the solution of integral equations a major issue is the efficient
evaluation of the complicated free-surface Green's functions and this is
addressed in §4.6.

Given the Green's function, the potential can then be represented in
the form of a source distribution, as in equation (4.8), and the boundary
condition on S_B then leads to an integral equation identical in form
to (4.12). Alternatively, an application of Green's theorem leads to
a representation and integral equation in the form of equations (4.36)

and (4.37). As noted by John (1950), failure of the solution of the integral-equation solution will occur at irregular values of the frequency parameter K corresponding to the eigenvalues of an interior Dirichlet problem as in §4.1.2. If S_B is completely submerged then the interior Dirichlet problem has no non-trivial solutions and the integral equation has no irregular values. If S_B intersects the free surface then there are irregular values of K corresponding to eigenvalues of the problem

$$\nabla^2 \phi(P_-) = 0, \quad P_- \in D_-, \tag{4.77}$$

$$\phi(p_-) = 0, \quad p_- \in S_B, \tag{4.78}$$

$$\frac{\partial \phi}{\partial n_p}(p_-) = K\phi(p_-), \quad p_- \in F_-, \tag{4.79}$$

where D_- and F_- are, respectively, the interior fluid domain and free surface. A modification of the Green's function to eliminate irregular values in a class of two-dimensional problems is given by Ursell (1981).

Unlike the problem discussed in §4.1, for some S_B finite-energy, non-trivial solutions of the homogeneous free-surface boundary-value problem (or 'trapped modes') are known to exist at specific values of K. Thus, for such S_B, the solution of the integral equation will fail at a trapped-mode frequency, as well as any irregular frequency. The discovery of trapped modes in the free-surface problem is quite recent (McIver 1996a) and extensive investigation of the phenomenon is still under way by a number of researchers. For recent progress on this topic see McIver (2000a).

Bibliographical notes

Because of their ability to handle complex geometries, computer codes based on integral-equation techniques have been widely adopted by the offshore industry for the solution of time-harmonic water-wave problems. Reviews of such numerical methods are given by Mei (1978) and Yeung (1982). Recent work on the elimination of irregular values in conventional integral-equation formulations includes Lau and Hearn (1989), Lee and Sclavounos (1989), Liapis (1992), Lee, Newman, and Zhu (1996).

For problems involving three-dimensional axisymmetric structures the integral equation (4.12), which involves an integral over the surface of the body, can be reduced to an infinite number of integral equations, each one involving only a line integral. Only two of these reduced integral equations need to be solved in order to compute the forces on the body. Details of this procedure can be found in Black (1975), Fenton

(1978), Isaacson (1982), Hulme (1983), and Hudspeth, Nakamura, and Pyun (1994). A extension to include terms of second order in the wave amplitude is given by Kim and Yue (1989).

Other methods have been proposed for the numerical solution of water-wave problems. The null-field method (Martin 1981, 1984), which involves constructing a bilinear expansion for the Green's function that leads to an infinite system of equations for the boundary-values of the unknown, does not suffer from the problem of irregular values. Neste-gard and Sclavounos (1984) and Eatock Taylor and Hu (1991) used an integral equation formulation near the structure and coupled this with a multipole expansion in the far field. Another approach followed by Yue, Chen, and Mei (1978) and Aranha, Mei, and Yue (1979) is to couple the finite-element method near the structure with an eigenfunction expansion in the far field. Both of these last two methods appear to be free of the problem of irregular values.

Although one of the great advantages of integral equation formulations over the techniques described in Chapters 2 and 3 is their applicability to a much wider class of geometries, the integral equations (4.12) and (4.37) can also be used as the starting point for analysis of wave interactions with structures which fit nicely into a particular coordinate system. Thus, for example, Wu and Eatock Taylor (1987, 1989) solved radiation and scattering problems for a submerged spheroid by expanding the unknown source distribution in (4.12) in spheroidal harmonics, an approach originally used by Farell (1973) for a different type of problem. Similarly, Gray (1978) studied the scattering of surface waves by a submerged sphere by expanding both the potential and the Green's function in (4.37) in terms of spherical harmonics.

4.6 Numerical evaluation of Green's functions

When solving boundary-value problems for the velocity potential exterior to a floating or submerged body via the integral equation approach described in §4.5 it is necessary to compute the appropriate Green's function, G, many times. For three-dimensional radiation and diffraction problems involving realistic body geometries the number of evaluations of G that need to be performed can be of the order of 10^6 for each frequency at which results are desired (Newman 1992). Furthermore, it may be necessary to analyze responses at up to 100 different frequencies in order to obtain a good description of the hydrodynamic characteristics of a particular structure. Clearly then, the efficient computation of such functions is desirable.

For engineering applications, the accuracy required in the final results is usually no more than three significant figures, but the computations involve taking spatial derivatives of the Green's function and inverting a large linear system of equations which is unlikely to be well conditioned. Thus it is not inappropriate to require G to be evaluated to an accuracy of six or seven decimals. Another reason for desiring very high accuracy in the determination of G is so that results can be obtained for standard problems which will serve as benchmarks against which new programmes and techniques can be measured.

Two different types of Green's function will be considered below. The first is the standard free-surface Green's function in three dimensions (both the infinite and finite depth cases are considered). This function is not that difficult to evaluate numerically, but in view of the fact that in practical applications it needs to be calculated millions of times, it is important to explore all the means by which its computation can be accelerated. The second type of Green's function that will be considered is that appropriate to problems involving obstacles in channels of finite width and depth. Such a Green's function might be used if trying to quantify the effects of the channel walls on the hydrodynamic characteristics of a body measured in a wave-tank experiment. This Green's function represents considerably more of a challenge when it comes to its efficient numerical calculation.

4.6.1 Green's functions for three-dimensional water-wave problems

For fully three-dimensional water-wave problems the Green's function is defined for infinite depth in §B.3 and for constant finite depth in §B.4. In each case the imaginary part of the function is given explicitly and so it is only the computation of the real part that is of interest. The infinite depth case will be considered first.

After the singular part of the Green's function has been subtracted off we are essentially left with the problem of computing a function of two variables $F(x, y)$ for which the three expressions (B.65), (B.69), and

(B.70) give, on associating $K(z + \zeta)$ with $-y$ and KR with x,

$$F(x,y) = \fint_0^\infty \frac{2}{u-1}\, \mathrm{e}^{-uy}\, J_0(ux)\, \mathrm{d}u, \tag{4.80}$$

$$F(x,y) = -2\pi\, \mathrm{e}^{-y}\, Y_0(x) - \frac{4}{\pi} \int_0^\infty \frac{\cos uy + u \sin uy}{u^2 + 1} K_0(ux)\, \mathrm{d}u, \tag{4.81}$$

$$F(x,y) = -\pi\, \mathrm{e}^{-y}\, (\mathbf{H}_0(x) + Y_0(x)) - 2 \int_0^y \frac{\mathrm{e}^{u-y}}{(x^2 + u^2)^{1/2}}\, \mathrm{d}u, \tag{4.82}$$

respectively. The Green's function is then given by

$$\frac{4\pi}{K}G = \frac{1}{Kr} + \frac{1}{Kr_1} + F(KR, K|z + \zeta|) + 2\pi\mathrm{i}\, \mathrm{e}^{K(z+\zeta)}\, J_0(KR). \tag{4.83}$$

We need to be able to evaluate $F(x,y)$ for all $y > 0$ and $x \geq 0$. The Bessel functions J_0 and Y_0 and the Struve function $\mathbf{H}_0$ can all be computed efficiently using standard routines; for example, Newman (1984a) provides polynomial approximations for J_0 and Y_0 and a rational approximation to $\mathbf{H}_0$, all with absolute errors less than 10^{-8}.

For typical parameter values that would be appropriate in ship-wave calculations, Hearn (1977) showed that computation of F from (4.82), which involves only standard functions and a finite non-singular integral, was in general significantly quicker than using (4.80) or (4.81), with (4.80) being more efficient than (4.81), particularly for small x.

For small x and large y, (4.80) is the most efficient starting point for computations. The Cauchy principal-value integral can be readily evaluated using one of the following techniques. First, since

$$\fint_0^2 (u-1)^{-1}\, \mathrm{d}u = 0, \tag{4.84}$$

it follows that

$$F(x,y) = \int_0^2 \frac{\mathrm{e}^{-uy}\, f(u) - \mathrm{e}^{-y}\, f(1)}{u - 1}\, \mathrm{d}u + \int_2^\infty \frac{f(u)}{u - 1}\, \mathrm{e}^{-uy}\, \mathrm{d}u, \tag{4.85}$$

where $f(u) = 2J_0(ux)$. This is often referred to as Monacella's method in the literature (Monacella 1966). Another method, used by Endo (1987) when calculating the finite depth Green's function, is to write

$$F(x,y) = \int_0^\infty \frac{f(u) - f(1)}{u - 1}\, \mathrm{e}^{-uy}\, \mathrm{d}u - f(1)\, \mathrm{e}^{-y}\, \mathrm{Ei}(y), \tag{4.86}$$

where Ei is the exponential integral (Abramowitz and Stegun 1965, §5.1). In both methods the integrand in the first integral is continuous at $u = 1$.

Alternatively, (4.80) can be used to derive a series expansion for F which is useful when x/y is small. If $J_0(ux)$ is expanded in powers of $(ux)^2$ (Abramowitz and Stegun 1965, eqn 9.1.10) and then the integral evaluated term by term one can obtain (see Noblesse 1982)

$$F(x,y) = 2 \sum_{n=0}^{\infty} \frac{(-x^2/4)^n}{(n!)^2} \left(\sum_{m=1}^{2n} \frac{(m-1)!}{y^m} - e^{-y} \operatorname{Ei}(y) \right). \quad (4.87)$$

According to Newman (1985), six decimal place accuracy can be obtained for all $x/y < 0.5$ by truncating this series at $n = 9$.

There are of course other ways of developing series expansions for F. One possibility is to expand the exponential in (4.82) in powers of u and integrate the resulting expression term by term. Details can be found in Newman (1984b). The resulting expression in terms of positive powers of both x and y is uniformly convergent throughout $x > 0$, $y > 0$. When x/y is large it is possible to derive a useful asymptotic series from (4.82) by expanding the reciprocal square root in the integrand in powers of $(u/x)^2$ and integrating term by term. It is also possible to produce approximations valid when both x and y are large (see Newman 1985).

It is thus clear that there are many different analytic expressions that can be used as starting points for the calculation of the Green's function and which is the most appropriate will depend critically on the given parameter values. Thus Telste and Noblesse (1986) presented a method based on a decomposition of the quadrant in which F is defined into five subregions and the use of different analytic representations in each of these regions. The numerical procedure advocated by Newman (1992) for the evaluation of F is also based on the subdivision of the quadrant $x > 0$, $y > 0$ into various subregions. The analytic expressions given above and others from the papers cited provide the necessary information to be able to separate out from the calculations any singularities and persistent oscillations so as to leave slowly-varying functions of two variables to be evaluated. These can then be computed using economized polynomial approximations.

Thus, given a well-behaved function $f(x_1, x_2)$ defined on $|x_1| \leq 1$, $|x_2| \leq 1$, we can assume an expansion in Chebyshev polynomials of the form

$$f(x_1, x_2) = \sum_{m=0}^{\infty} \sum_{n=0}^{\infty} c_{mn} T_m(x_1) T_n(x_2). \quad (4.88)$$

The coefficients c_{mn} are easily computed using orthogonality relations for the Chebyshev polynomials and can then be stored. In the actual implementation reported by Newman, which is designed to produce six decimal place accuracy over the whole domain $x > 0$, $y > 0$, the quadrant is subdivided into 48 subdomains and 31 polynomial coefficients are needed in each of these domains.

For the finite depth case the Greens's function is given in terms of a Cauchy principal-value integral by (B.86), but there is no representation in terms of finite integrals analogous to (4.82). However there is now an eigenfunction expansion (B.91). This expression converges rapidly provided R/h is large enough and for $R/h \geq 0.5$ a maximum of 12 terms are required to achieve an accuracy of six decimals. (For larger values of R/h, $[6h/R]$ terms are sufficient for this level of accuracy.) For small R/h the eigenfunction expansion is unsuitable for computations since the individual terms become singular in the limit as $R/h \to 0$ and for $R/h < 0.5$, Newman (1992) describes algorithms based on triple economized polynomial approximations to functions of three variables, requiring a total number of stored coefficients of about 8000.

An alternative approach to the computation of the finite depth Green's function has been developed by Linton (1999b). If we write

$$Z_m = \frac{1}{hN_m^2} \cos k_m(z + h) \cos k_m(\zeta + h), \qquad (4.89)$$

then the real part of the Green's function is g where

$$4\pi g = -\pi Z_0 Y_0(kR) + 2 \sum_{m=1}^{\infty} Z_m K_0(k_m R) \qquad (4.90)$$

$$= -\pi Z_0 Y_0(kR) + \sum_{m=1}^{\infty} Z_m \int_0^{\infty} e^{-R^2/4t} e^{-k_m^2 t} \frac{dt}{t}, \qquad (4.91)$$

where the integral representation for the modified Bessel function K_0 given by Gradshteyn and Ryzhik (1980 eqn 8.432(7)) has been used. The procedure now is to split the integral at an arbitrary point, a^2 say, and then treat the two resulting integrals differently. By varying a the convergence characteristics of the resulting expressions can be altered. If we define

$$\Lambda_0 = -\pi Y_0(kR) - \int_0^{a^2} e^{-R^2/4t} e^{k^2 t} \frac{dt}{t}, \qquad (4.92)$$

$$\Lambda_m = \int_{a^2}^{\infty} e^{-R^2/4t} e^{-k_m^2 t} \frac{dt}{t}, \qquad m \geq 1, \qquad (4.93)$$

then (4.91) becomes

$$4\pi g = \sum_{m=0}^{\infty} \Lambda_m Z_m + \int_0^{a^2} e^{-R^2/4t} \, w(t) \frac{dt}{t}, \qquad (4.94)$$

where

$$w(t) = \sum_{m=0}^{\infty} Z_m \, e^{-k_m^2 t}. \qquad (4.95)$$

It is straightforward to evaluate the integrals Λ_m, $m \geq 1$, numerically due to the monotonic exponential decay of the integrand and the same is true of Λ_0, except when R is very small. In the latter case it is possible to derive a series expansion by expanding $\exp(k^2 t)$ in powers of t and integrating term by term. Note that Λ_0 is not singular as $R \to 0$; in fact, when $R = 0$,

$$\Lambda_0 = -\operatorname{Ei}\left(k^2 a^2\right), \qquad \Lambda_m = E_1\left(k_m^2 a^2\right), \qquad (4.96)$$

where Ei and E_1 are exponential integrals as defined in Abramowitz and Stegun (1965, §5.1).

The expression for w converges rapidly for large values of t but the integral in (4.94) is over values of t near the origin so in order to develop an expression for the Green's function which is easy to compute we require a representation of w which converges rapidly for small t. Such an expansion was derived in Linton (1999b) and if this is substituted into (4.94) and some of the integrals evaluated we obtain

$$\begin{aligned}
4\pi g = \sum_{m=0}^{\infty} \Lambda_m Z_m &+ \frac{1}{r} \operatorname{erfc}\left(\frac{r}{2a}\right) + \frac{1}{r_2} \operatorname{erfc}\left(\frac{r_2}{2a}\right) \\
&+ \sum_{i=1}^{4} \frac{1}{(R^2 + \chi_i^2)^{1/2}} \operatorname{erfc}\left(\frac{(R^2 + \chi_i^2)^{1/2}}{2a}\right) \\
&+ 2K \int_0^a \left[e^{K^2 u^2 - R^2/4u^2} \sum_{i=1}^{4} e^{-K\chi_i} \operatorname{erfc}\left(\frac{\chi_i}{2u} - Ku\right) + Q \right] \frac{du}{u},
\end{aligned} \qquad (4.97)$$

in which

$$\chi_1 = -\zeta - z, \qquad \chi_2 = 2h - \zeta + z,$$
$$\chi_3 = 2h + \zeta - z, \qquad \chi_4 = 4h + \zeta + z$$

and erfc is the complementary error function (Abramowitz and Stegun 1965, §7.1). The term Q (which is actually an infinite series of complicated expressions) can be omitted from numerical calculations as will be described below.

The value of g is independent of a, but this parameter crucially affects the convergence of the infinite series in (4.97). It also affects the magnitude of the error introduced by omitting Q from the calculations and this error (for which a simple bound can be obtained) can be made smaller than the desired accuracy for g by using a small enough value for a. We note that in the limit as $a \to 0$ we recover the eigenfunction expansion (4.90) which, as has already been noted, is numerically efficient when R/h is large. For small values of R/h we can use (4.97) with a non-zero value of a. The larger a is the fewer terms are required in the infinite series, but the error introduced by omitting Q from the calculations puts an upper bound on the value of a that can be used. It was shown in Linton (1999b) that this error is less than 10^{-8} when $a = h/4$. By this process values of g accurate to six places of decimals can be obtained for all R by truncating the infinite series in (4.97) at $m = 3$. Of particular note is that when $|z + \zeta|$ is very small and $R = 0$, a situation in which this particular Green's function is otherwise very difficult to evaluate, the representation (4.97) takes a particularly simple form due to (4.96) and still converges rapidly.

4.6.2   Channel Green's functions

Consider a water-wave channel of width d and depth h. For problems in which the depth dependence can be factored out (such as scattering by bottom-mounted surface-piercing cylinders of constant cross-section) the Green's function $G(x, y; \xi, \eta)$ satisfies the two-dimensional Helmholtz equation (4.1), with a logarithmic singularity as (x, y) approaches (ξ, η), and the wall boundary conditions

$$\frac{\partial G}{\partial y} = 0, \quad \text{on} \quad y = 0, d. \tag{4.98}$$

This is exactly equivalent to seeking the Green's function for a two-dimensional acoustic wave guide with rigid walls. This Green's function can be written in terms of a function of two variables as

$$G(x, y, \xi, \eta) = \widetilde{G}(X, y - \eta) + \widetilde{G}(X, y + \eta), \tag{4.99}$$

where $X = x - \xi$ and $\widetilde{G}$ represents an infinite row of line sources at $(x, y) = (0, 2md)$, $m = 0, \pm 1, \pm 2, \dots$. There are many alternative representations for $\widetilde{G}$ (see Linton 1998 and the references cited therein), perhaps the best known being the image series

$$\widetilde{G}(x,y) = -\frac{\mathrm{i}}{4} \sum_{m=-\infty}^{\infty} H_0^{(1)}(kr_m), \tag{4.100}$$

where $r_m = [x^2 + (y - 2md)^2]^{1/2}$, and the eigenfunction expansion

$$\widetilde{G}(x,y) = -\frac{1}{4} \sum_{m=0}^{\infty} \frac{\epsilon_m}{\alpha_m} \, \mathrm{e}^{-\alpha_m |x|/d} \cos \frac{m\pi y}{d}, \tag{4.101}$$

where

$$\alpha_m = \left(m^2\pi^2 - k^2 d^2\right)^{1/2} = -\mathrm{i}\left(k^2 d^2 - m^2\pi^2\right)^{1/2} \tag{4.102}$$

and ϵ_m is the Neumann symbol defined on page 17.

The image series represents a sum of sources (each given by equation 4.4) arranged periodically along the y-axis and clearly shows, see (A.17), that as $kr_0 \to 0$, $\widetilde{G} \sim \ln r_0$. However, it is useless from a computational point of view due to the slow convergence of the summation. The eigenfunction expansion on the other hand does not display the singularity at the origin explicitly, but has the advantage that the imaginary part of $\widetilde{G}$ can immediately be obtained as a finite series, the only contributions coming from those values of m for which α_m is imaginary. The series is also useful for numerical computations when $|x|$ is large, but converges extremely slowly when $x = 0$. Various techniques for accelerating this series are collected together in Linton (1998).

Another approach to the computation of $\widetilde{G}$ is to use an integral representation. One such representation, which appears to be computationally efficient, is

$$4\mathrm{i}\,\widetilde{G}(x,y) = H_0^{(1)}(kr_0) - \frac{4\mathrm{i}}{\pi} \int_0^{\infty} \frac{\cosh[ky(u-\mathrm{i})]\cos[kxf(u)]}{(\mathrm{e}^{2kd(u-\mathrm{i})}-1)f(u)}\,\mathrm{d}u, \tag{4.103}$$

where $f(u) = (u^2 - 2\mathrm{i}u)^{1/2}$.

In applications where a large number of computations of $\widetilde{G}$ are required for each value of kd, the so-called lattice sum technique can be competitive. If Graf's addition theorem for Bessel functions (Gradshteyn and Ryzhik 1980, eqn 8.530(2)) is applied to the image series (4.100) we obtain

$$4\mathrm{i}\,\widetilde{G}(x,y) = H_0^{(1)}(kr_0) + \sum_{\ell=0}^{\infty} \epsilon_\ell (-1)^\ell S_{2\ell}(2kd) J_{2\ell}(kr_0) \cos 2\ell\theta, \quad (4.104)$$

where $x = r_0 \cos\theta$, $y = r_0 \sin\theta$ and

$$S_n(t) = 2 \sum_{m=1}^{\infty} H_n^{(1)}(mt). \quad (4.105)$$

The convergence of the sum over ℓ is very rapid and the main computational effort comes from the evaluation of the lattice sums $S_{2\ell}(2kd)$. Representations for these sums in terms of series which converge reasonably quickly were given by Twersky (1961) (see also Miles 1983, Appendix A) and accelerated versions are given in Linton (1998). Since the lattice sums do not depend on either x or y they only have to be computed once for each value of kd, hence a considerable computational saving can be achieved if many spatial evaluations are required for each frequency of interest.

Finally, a representation for $\widetilde{G}$ can be derived in terms of an arbitrary parameter a, much as for the finite depth water-wave Green's function above, though in this case the result is somewhat simpler:

$$\widetilde{G} = -\frac{1}{8} \sum_{m=0}^{\infty} \frac{\epsilon_m}{\alpha_m} \cos\frac{m\pi y}{d} \left[f_m^+(x) + f_m^-(x) \right]$$
$$-\frac{1}{4\pi} \int_0^{a^2 d^2} \left(\sum_{m=-\infty}^{\infty} \mathrm{e}^{-r_m^2/4t} \right) \mathrm{e}^{k^2 t}\,\frac{dt}{t}, \quad (4.106)$$

where

$$f_m^\pm(x) = \mathrm{e}^{\pm\alpha_m x/d} \operatorname{erfc}\left(\alpha_m a \pm \frac{x}{2ad}\right). \quad (4.107)$$

The integral in (4.106) is easily evaluated numerically due to the monotonic and exponential decay of the integrand. For small values of r_0, the $m = 0$ term from the second summation is best treated separately. If $\exp(k^2 t)$ is expanded in powers of t and the result integrated term by term we obtain

$$\int_0^{a^2 d^2} \mathrm{e}^{-r_0^2/4t}\,\mathrm{e}^{k^2 t}\,\frac{dt}{t} = \sum_{n=0}^{\infty} \frac{(kad)^{2n}}{n!} E_{n+1}\left(\frac{r_0^2}{4a^2 d^2}\right), \quad (4.108)$$

from which we recover the fact that

$$\widetilde{G} \sim -\frac{1}{4\pi} E_1 \left(\frac{r_0^2}{4a^2 d^2} \right) \sim \frac{1}{2\pi} \ln r_0 \quad \text{as} \quad r_0 \to 0. \tag{4.109}$$

In the limit as $a \to 0$ we recover the eigenfunction expansion (4.101) from (4.106). By increasing a we can change the relative importance of the two summations in (4.106) and by balancing the convergence rates an extremely efficient method for the numerical evaluation of $\widetilde{G}$ can be developed. Typically, values of a in the range $0.2 < a < 0.5$ give the best results.

For channel problems where the depth dependence cannot be factored out a computationally efficient representation for the Green's function can be derived by combining (4.97) with (4.106). The details can be found in Linton (1999a).

4.7 Diffraction by a breakwater

To illustrate the power of integral equation methods, the problem of diffraction of waves by a breakwater is now considered in some detail. A straight, rigid, vertical breakwater stands in water of uniform depth. The thickness of the breakwater is assumed to be negligible and Cartesian coordinates are chosen so that the x,y-plane is horizontal and the breakwater coincides with the x-axis. Standard plane polar coordinates (r, θ) defined by $x = r \cos \theta$ and $y = r \sin \theta$ will also be used.

Two configurations of breakwater will be considered. In the first the breakwater is of infinite extent but contains a single gap, and in the second the breakwater has finite length.

4.7.1 Diffraction by a gap in a breakwater

In this section, the breakwater occupies that part of the y-axis for which $|x| \geq a$ so that it has a single gap in $|x| < a$; for convenience this last interval will be denoted by L. A wave with wavenumber k is incident in $y > 0$ at an angle β to the x-axis (see Figure 4.1); from the symmetry of the problem it is sufficient to consider $\beta \in [0, \pi/2]$. The initial aim is to derive an integral equation that may be used to determine the wave field resulting from the diffraction of this wave by the gap. This problem

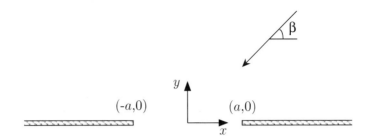

FIGURE 4.1
Sketch of the breakwater-gap problem.

is equivalent to one in two-dimensional acoustics in which an incident
wave is diffracted by an aperture in a rigid screen.

The problem may be solved in terms of a velocity potential

$$\Phi_T(x, y, t) = \mathrm{Re}\left\{\phi_T(x, y)\, e^{-i\omega t}\right\}, \qquad (4.110)$$

where ϕ_T, for the reasons explained in §2.4, satisfies the two-dimensional
Helmholtz equation (2.74), has zero normal derivative on the breakwater,
and satisfies a radiation condition specifying that the diffracted waves
are outgoing. The incident wave has a potential

$$\phi_I = e^{-ikr\cos(\theta-\beta)} \qquad (4.111)$$

and it is convenient to extract from the total potential ϕ_T both ϕ_I and
its reflection from a rigid barrier at $y = 0$. Thus, write

$$\phi_T(x, y) = \begin{cases} 2\cos(ky\sin\beta)\, e^{-ikx\cos\beta} +\phi^+(x, y), & y > 0, \\ \phi^-(x, y), & y < 0, \end{cases} \qquad (4.112)$$

so that the potentials $\phi^\pm$ represent the diffraction by the gap and satisfy
the Helmholtz equation and the radiation conditions

$$\lim_{kr\to\infty} r^{1/2}\left(\frac{\partial\phi^\pm}{\partial r} - ik\phi^\pm\right) = 0 \qquad (4.113)$$

in the appropriate half plane. To ensure uniqueness of the solution the
edge conditions that ϕ_T is bounded at the breakwater tips $(x, y) =
(\pm a, 0)$ are imposed. This leads to a solution satisfying

$$|\nabla\phi_T| = O(r_\pm^{-1/2}) \quad \text{as} \quad r_\pm \to 0, \qquad (4.114)$$

where $r_\pm$ denote the distances from the edges $(x, y) = (\pm a, 0)$, so that the energy of the fluid motion is bounded in all compact subsets of the fluid domain. For further discussion of the edge condition see Jones (1986, §9.2).

With the above construction, the condition of no flow through the breakwater requires that

$$\frac{\partial \phi^+}{\partial y}(x, 0) = \frac{\partial \phi^-}{\partial y}(x, 0) = 0, \quad x \notin L. \tag{4.115}$$

Further, there must be continuity of potential and velocity in the gap which give, respectively,

$$2 e^{-ikx \cos \beta} + \phi^+(x, 0) = \phi^-(x, 0), \quad x \in L, \tag{4.116}$$

and

$$\frac{\partial \phi^+}{\partial y}(x, 0) = \frac{\partial \phi^-}{\partial y}(x, 0) \equiv v_\beta(x), \quad x \in L. \tag{4.117}$$

An integral equation for $v_\beta(x)$ is now derived by applications of Green's theorem (4.36) to $\phi^\pm$ and the Green's function

$$G(x, y; \xi, \eta) = \frac{1}{4i}\left[H_0^{(1)}(kR) + H_0^{(1)}(kR')\right], \tag{4.118}$$

where

$$R = \left[(x - \xi)^2 + (y - \eta)^2\right]^{1/2} \tag{4.119}$$

is the distance between the field point (x, y) and the source point (ξ, η), and

$$R' = \left[(x - \xi)^2 + (y + \eta)^2\right]^{1/2} \tag{4.120}$$

is the distance between the field point and the image point $(\xi, -\eta)$; by construction

$$\frac{\partial G}{\partial y}(x, 0; \xi, \eta) = 0. \tag{4.121}$$

First of all the contour of integration is chosen as $y = 0^+$ together with an enclosing semicircle in $y > 0$ (see Figure 4.2). Green's theorem (4.36) yields

$$\phi^+(\xi, \eta) = -\frac{i}{2} \int_L v_\beta(x) H_0^{(1)}(kR)\, dx, \quad \eta > 0, \tag{4.122}$$

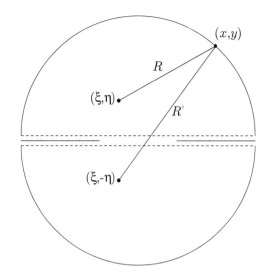

FIGURE 4.2
Contours for the application of Green's theorem to the break-water gap problem.

where the boundary condition (4.115), the continuity condition (4.117), and also (4.121) and the fact that $R = R'$ on $y = 0$ have been used to simplify the integrations. A similar calculation with the contour of integration chosen as $y = 0^-$ and an enclosing semicircle in $y < 0$ gives

$$\phi^-(\xi, -\eta) = \frac{\mathrm{i}}{2} \int_L v_\beta(x) H_0^{(1)}(kR)\, \mathrm{d}x, \quad \eta > 0. \tag{4.123}$$

The last two equations are representations for ϕ in terms of the unknown velocity $v_\beta(x)$ and it follows immediately that

$$\phi^-(x, -y) = -\phi^+(x, y), \quad y \geq 0. \tag{4.124}$$

An integral equation for $v_\beta(x)$ is obtained by using the representations (4.122) and (4.123) in the the potential continuity condition (4.116) to obtain

$$\int_L h(|x - t|) v_\beta(t)\, \mathrm{d}t = f_\beta(x), \quad x \in L, \tag{4.125}$$

where

$$h(x) = \frac{\mathrm{i}}{2} H_0^{(1)}(kx) \quad \text{and} \quad f_\beta(x) = \mathrm{e}^{-\mathrm{i}kx \cos \beta}. \tag{4.126}$$

The integral equation has been written in the form (4.125) to empha-
size that some of the results to follow in §4.7.3 are not restricted to
the breakwater-gap problem, but apply to a general class of integral
equations with difference kernels.

An important physical quantity is the amplitude of the diffracted
waves in the far field. This is readily obtained from (4.122) and (4.123)
using the asymptotic behaviour of the Hankel function given in equation
(A.12). A change of notation in (4.122) gives

$$\phi^+(x, y) = -\frac{i}{2} \int_L v_\beta(t) H_0^{(1)}(k\rho) \, dt, \quad y > 0, \tag{4.127}$$

where

$$\rho = (r^2 + t^2 - 2rt \cos \theta)^{1/2} \sim r - t \cos \theta \quad \text{as} \quad r/t \to \infty \tag{4.128}$$

and so for $\theta \in (0, \pi)$

$$\phi^+(x, y) \sim \frac{e^{i(kr - 3\pi/4)}}{(2\pi kr)^{1/2}} G(\theta, \beta) \quad \text{as} \quad kr \to \infty, \tag{4.129}$$

where

$$G(\theta, \beta) = \int_L v_\beta(t) \, e^{-ikt \cos \theta} \, dt = \langle v_\beta, f_{\pi - \theta} \rangle \tag{4.130}$$

is known as the diffraction coefficient and the inner product is defined
by

$$\langle v, w \rangle \equiv \int_L v(t) \overline{w(t)} \, dt. \tag{4.131}$$

The diffracted field in the lower half plane follows from (4.124).

From the definition (4.130) and the integral equation (4.125)

$$G(\theta, \beta) = \int_L v_\beta(x) f_\theta(x) \, dx = \int_L v_\beta(x) \left\{ \int_L h(|x - t|) v_\theta(t) \, dt \right\} dx$$

$$= \int_L v_\theta(t) \left\{ \int_L h(|x - t|) v_\beta(x) \, dx \right\} dt$$

$$= \int_L v_\theta(t) f_\beta(t) \, dt = G(\beta, \theta). \tag{4.132}$$

This reciprocity principle says that the far-field diffracted waves measured in the direction θ due to waves incident at an angle β are the same
as those observed at an angle β due to waves incident at an angle θ.

4.7.2   Diffraction by an insular breakwater

A closely-related problem to that of diffraction by a breakwater gap is that of diffraction by an 'insular' breakwater, that is, a straight, rigid, vertical breakwater of finite length. The problem is similar to that described in §4.7.1 except that the breakwater occupies $|x| \leq a$ instead of $|x| \geq a$. This problem is equivalent to one in two-dimensional acoustics in which an incident wave is diffracted by a rigid ribbon.

The incident wave is removed from the total potential throughout the fluid domain so that

$$\phi_T(x, y) = e^{-ikr\cos(\theta - \beta)} + \phi(x, y), \tag{4.133}$$

where the diffraction potential ϕ satisfies the Helmholtz equation, the condition of no flow through the breakwater,

$$\frac{\partial \phi}{\partial y}(x, 0) = ik \sin \beta \, e^{-ikx\cos\beta}, \quad x \in L, \tag{4.134}$$

the radiation condition

$$\lim_{kr\to\infty} r^{1/2} \left(\frac{\partial \phi}{\partial r} - ik\phi \right) = 0 \quad \text{as} \quad r \to \infty, \tag{4.135}$$

and an edge condition similar to (4.114).

A suitable integral equation for the problem is derived in a very similar way to that used in §4.7.1. First of all, applications of Green's theorem (4.36) in the upper and lower half planes to ϕ and the Green's function defined in (4.118) yield

$$\phi(\xi, \pm\eta) = \mp \frac{i}{2} \int_{-\infty}^{\infty} \frac{\partial \phi}{\partial y}(x, 0^{\pm}) H_0^{(1)}(kR) \, dx, \quad \eta > 0, \tag{4.136}$$

where $y = 0$. Hence, from the continuity of $\partial \phi/\partial y$ on $y = 0$,

$$\phi(x, -y) = -\phi(x, y) \tag{4.137}$$

and, as ϕ is continuous for $|x| > a$, it then follows that

$$\phi(x, 0) = 0, \quad |x| > a. \tag{4.138}$$

If instead Green's theorem (4.36) is applied in the upper half plane to ϕ and the Green's function

$$G(x, y; \xi, \eta) = \frac{1}{4i} \left[H_0^{(1)}(kR) - H_0^{(1)}(kR') \right], \tag{4.139}$$

which satisfies

$$G(x, 0; \xi, \eta) = 0 \qquad (4.140)$$

and with the aid of

$$\frac{\partial H_0^{(1)}}{\partial y}(kR') = -\frac{\partial H_0^{(1)}}{\partial y}(kR) \quad \text{on} \quad y = 0, \qquad (4.141)$$

this results in the alternative representation

$$\phi(\xi, \eta) = \frac{i}{2} \int_L \phi(x, 0^+) \frac{\partial H_0^{(1)}}{\partial y}(kR)\, dx, \quad \eta > 0, \qquad (4.142)$$

where $y = 0$, and (4.138) has been used to reduce the range of integration. It remains to satisfy the boundary condition (4.134). Now

$$\frac{\partial H_0^{(1)}}{\partial y}(kR) = -\frac{\partial H_0^{(1)}}{\partial \eta}(kR) \qquad (4.143)$$

so that

$$\frac{\partial \phi}{\partial \eta}(\xi, 0) = \frac{i}{2} \frac{\partial}{\partial \eta} \int_L \phi(x, 0^+) \frac{\partial H_0^{(1)}}{\partial y}(kR)\, dx \bigg|_{\eta=0}$$

$$= -\frac{i}{2} \frac{\partial^2}{\partial \eta^2} \int_L \phi(x, 0^+) H_0^{(1)}(kR)\, dx \bigg|_{\eta=0}$$

$$= \frac{i}{2} \left(\frac{\partial^2}{\partial \xi^2} + k^2 \right) \int_L \phi(x, 0^+) H_0^{(1)}(k|x - \xi|)\, dx, \qquad (4.144)$$

and hence, with a change of notation, (4.134) yields the integral equation

$$\left(\frac{d^2}{dx^2} + k^2 \right) \int_L h(|x - t|)\phi_\beta(t)\, dt = ik \sin \beta\, f_\beta(x), \quad x \in L, \qquad (4.145)$$

for the potential $\phi_\beta(x) \equiv \phi(x, 0^+)$. From (4.138) this potential satisfies the boundary conditions

$$\phi_\beta(\pm a) = 0. \qquad (4.146)$$

The far-field diffracted waves are calculated in a similar way to the gap problem in §4.7.1. It is found that

$$\phi(x, y) \sim \frac{e^{i(kr - 3\pi/4)}}{(2\pi kr)^{1/2}} F(\theta, \beta) \quad \text{as} \quad kr \to \infty, \tag{4.147}$$

where the diffraction coefficient

$$F(\theta, \beta) = ik \sin \theta \, \langle \phi_\beta, f_{\pi - \theta} \rangle \tag{4.148}$$

and the inner product notation is defined in equation (4.131).

4.7.3 Embedding formulas

It is a remarkable fact that the solutions of the integral equations (4.125) and (4.145) for all angles of wave incidence β can be expressed in terms of the solution of (4.125) for a single β. These results, known as embedding formulas, will now be derived. The formulas apply for general weakly singular kernels $h(|x - t|)$ as long as the corresponding integral operator is injective, as is the case for the Hankel-function kernel in the breakwater-gap problem.

First of all it is noted that the geometrical symmetry in the two problems immediately implies that

$$v_{\pi - \beta}(-x) = v_\beta(x) \quad \text{and} \quad \phi_{\pi - \beta}(-x) = \phi_\beta(x). \tag{4.149}$$

In turn, these imply that the diffraction coefficient

$$G(\pi - \theta, \pi - \beta) = \int_L v_{\pi - \beta}(x) \overline{f_\theta(x)} \, dx = \int_L v_{\pi - \beta}(-x) \overline{f_\theta(-x)} \, dx$$

$$= \int_L v_\beta(x) \overline{f_{\pi - \theta}(x)} \, dx = G(\theta, \beta) \tag{4.150}$$

and similarly

$$F(\pi - \theta, \pi - \beta) = F(\theta, \beta). \tag{4.151}$$

For convenience the integral equations (4.125) and (4.145) are rewritten using operator notation as

$$(Hv_\beta)(x) = f_\beta(x), \quad x \in L, \tag{4.152}$$

and

$$\left(\frac{\mathrm{d}^2}{\mathrm{d}x^2} + k^2\right)(H\phi_\beta)(x) = \mathrm{i}k\sin\beta\, f_\beta(x), \quad x \in L, \tag{4.153}$$

respectively. Integration by parts and application of $\phi_\beta(\pm a) = 0$ give

$$\frac{\mathrm{d}}{\mathrm{d}x}(H\phi_\beta)(x) = (H\phi'_\beta)(x) \tag{4.154}$$

so that (4.153) may be rewritten as

$$\left(\frac{\mathrm{d}^2}{\mathrm{d}x^2} + k^2\right)(H\phi_\beta)(x) = \left(\frac{\mathrm{d}}{\mathrm{d}x} + \mathrm{i}k\right)\left(\frac{\mathrm{d}}{\mathrm{d}x} - \mathrm{i}k\right)(H\phi_\beta)(x)$$

$$= \left(\frac{\mathrm{d}}{\mathrm{d}x} + \mathrm{i}k\right)\left\{H(\phi'_\beta - \mathrm{i}k\phi_\beta)\right\}(x) = \mathrm{i}k\sin\beta\, f_\beta(x), \quad x \in L. \tag{4.155}$$

The last equality is a first-order differential equation which has general solution

$$\left\{H(\phi'_\beta - \mathrm{i}k\phi_\beta)\right\}(x) = \cot\tfrac{1}{2}\beta\ (f_\beta(x) - c_\beta f_0(x)), \quad x \in L, \tag{4.156}$$

where c_β is the constant of integration, and it immediately follows from (4.152) and the assumption that H is injective that

$$\phi'_\beta(x) - \mathrm{i}k\phi_\beta(x) = \cot\tfrac{1}{2}\beta\ (v_\beta(x) - c_\beta v_0(x)), \quad x \in L. \tag{4.157}$$

The constant c_β is determined by taking the inner product with f_π of both sides of (4.157). Integration by parts and application of $\phi_\beta(\pm a) = 0$ show that

$$\langle \phi'_\beta - \mathrm{i}k\phi_\beta, f_\pi \rangle = 0 \tag{4.158}$$

and hence

$$c_\beta = G(\beta, 0)/G(0, 0) \tag{4.159}$$

so that

$$\phi'_\beta(x) - \mathrm{i}k\phi_\beta(x) = \cot\tfrac{1}{2}\beta \left\{v_\beta(x) - \frac{G(\beta, 0)}{G(0, 0)}v_0(x)\right\}, \quad x \in L. \tag{4.160}$$

If, in this equation, β is replaced by $\pi-\beta$ and x by $-x$ and the symmetry properties of ϕ_β and v_β used, this results in

$$-\phi'_\beta(x) - \mathrm{i}k\phi_\beta(x) = \tan\tfrac{1}{2}\beta \left\{ v_\beta(x) - \frac{G(\pi-\beta,0)}{G(0,0)} v_0(-x) \right\}, \quad x \in L. \tag{4.161}$$

Elimination of v_β from the last two equations gives the differential equation

$$\phi'_\beta(x) + \mathrm{i}k\cos\beta \ \phi_\beta(x) =$$
$$- \frac{\sin\beta}{2G(0,0)} \left\{ G(\beta,0)v_0(x) - G(\pi-\beta,0)v_0(-x) \right\}, \quad x \in L, \tag{4.162}$$

which has the solution

$$\phi_\beta(x) = \frac{f_\beta(x)\sin\beta}{2G(0,0)}$$
$$\times \int_x^a \left\{ G(\beta,0)v_0(t) - G(\pi-\beta,0)v_0(-t) \right\} f_{\pi-\beta}(t)\, \mathrm{d}t \tag{4.163}$$

satisfying $\phi(\pm a) = 0$. For any angle of incidence β, this expresses the solution to the insular breakwater problem entirely in terms of the solution to the breakwater-gap problem for angle of incidence zero. With ϕ_β determined, v_β follows from either of (4.160) or (4.161).

It is now a straightforward matter to determine expressions for the diffraction coefficients entirely in terms of the diffraction coefficients for the breakwater-gap problem for angle of incidence zero. The results are

$$F(\theta,\beta) = \sin\theta \ \sin\beta \left\{ \frac{G(\pi-\theta,0)G(\pi-\beta,0) - G(\theta,0)G(\beta,0)}{2G(0,0)(\cos\theta + \cos\beta)} \right\} \tag{4.164}$$

for the insular breakwater and

$$G(\theta,\beta) = \frac{\hat{G}(\theta)\hat{G}(\beta) - \hat{G}(\pi-\theta)\hat{G}(\pi-\beta)}{2G(0,0)(\cos\theta + \cos\beta)} \tag{4.165}$$

for the breakwater with gap, where

$$\hat{G}(\theta) = (1+\cos\theta)G(\theta,0). \tag{4.166}$$

The results in (4.164) and (4.165) are undefined for $\theta = \pi - \beta$, but the appropriate limiting forms can be obtained from L'Hôpital's rule. For example,

$$G(\pi - \beta, \beta) = -\frac{\hat{G}'(\pi - \beta)\hat{G}(\beta) + \hat{G}'(\beta)\hat{G}(\pi - \beta)}{2G(0,0)\sin\beta}. \qquad (4.167)$$

This expression is itself undefined for $\sin\beta = 0$ and $G(\pi, 0) = G(0, \pi)$ must be computed directly.

4.7.4 Numerical solutions

Here two numerical methods are presented for the solution of the integral equation for the breakwater-gap problem. In view of the embedding formulas in §4.7.3, at each frequency it is required only to solve the gap problem for zero angle of incidence in order to obtain the solutions for all angles of incidence for both the gap and insular breakwater problems. The solution for the insular breakwater may be obtained directly using the method given in §4.3.

First of all, the integral equation (4.125) is solved by a collocation method. In this method the integral is approximated by a summation and equations generated for values of $v_\beta(x)$ at discrete points within L. Before doing this it is advantageous to make a change of integration variable in view of the fact that $v_\beta(x) \sim C^{\pm}(a \mp x)^{-1/2}$ as $x \to \pm a$ (MacCamy 1958b). An appropriate change is to write

$$x = a\cos u \quad \text{and} \quad t = a\cos v \qquad (4.168)$$

so that (4.125) becomes

$$\frac{i}{2}\int_0^\pi H_0^{(1)}(ka|\cos u - \cos v|)\,V_\beta(v)\,\mathrm{d}v = e^{-ika\cos u \cos\beta}, \quad u \in (0, \pi), \tag{4.169}$$

where, by construction,

$$V_\beta(u) = a\,v_\beta(a\cos u)\sin u \qquad (4.170)$$

is non-singular for all $u \in [0, \pi]$. Divide the interval $[0, \pi]$ into N equal elements with end points $\{u_i = i\pi/N;\ i = 0, 1, \ldots, N\}$ and midpoints $\{\hat{u}_i = (u_{i-1} + u_i)/2;\ i = 1, 2, \ldots, N\}$. The integral equation is reduced to a set of simultaneous equations for $\{V_\beta(\hat{u}_i);\ i = 1, 2, \ldots, N\}$

		N			
ka	20	40	60	80	100
π	3.9953	3.9964	3.9965	3.9966	3.9966
2π	5.7076	5.7147	5.7154	5.7155	5.7156
3π	7.1095	7.1309	7.1328	7.1332	7.1334
4π	8.3918	8.4414	8.4452	8.4461	8.4464

Table 4.1 The modulus of the diffraction coefficient $G(11\pi/12, \pi/12)$ for various wavenumbers ka and numbers of collocation points N.

by approximating the integral by a summation and collocating at the midpoints of the elements to get

$$\frac{\mathrm{i}}{2} \sum_{j=1}^{N} V_\beta(\hat{u}_j) \int_{u_{j-1}}^{u_j} H_0^{(1)}(ka|\cos \hat{u}_i - \cos v|)\, \mathrm{d}v$$

$$= \mathrm{e}^{-\mathrm{i}ka \cos \hat{u}_i \cos \beta}, \quad i = 1, 2, \ldots, N. \quad (4.171)$$

This is a set of linear simultaneous equations that is readily solved for the velocities $\{V_\beta(\hat{u}_i),\ i = 1, 2, \ldots, N\}$. The diffraction coefficient then follows from (4.130) and is

$$G(\theta, \beta) \approx \sum_{j=1}^{N} V_\beta(\hat{u}_j) \int_{u_{j-1}}^{u_j} \mathrm{e}^{-\mathrm{i}ka \cos v \cos \theta}\, \mathrm{d}v. \quad (4.172)$$

The integrals in the last two equations are straightforward to calculate numerically by standard techniques, although it should be noted that the integrals in (4.171) contain a logarithmic singularity when $i = j$.

The rate of convergence of the numerical scheme depends on ka and the angles of incidence and observation. The convergence becomes slower with increasing wavenumber ka due to more rapid variations in V_β and results for various ka and increasing N are presented in Table 4.1 for the diffraction coefficient $G(\theta, \beta) \equiv G(11\pi/12, \pi/12)$. The particular angles of incidence and observation have been chosen because there is also a decrease in the rate of convergence near the lines $\theta = \pi - \beta$ as the angle of incidence β is decreased. Thus, the results illustrate what are roughly the slowest rates of convergence that can be expected from the numerical scheme. They are compared with those from a different

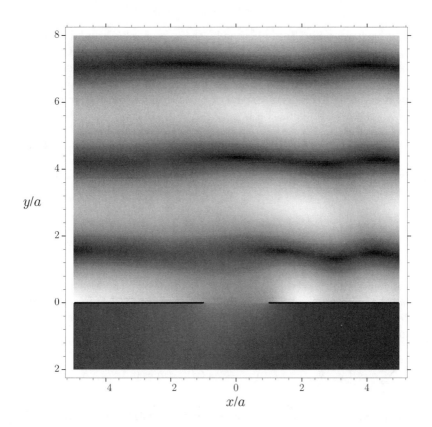

FIGURE 4.3
The wave field $|\phi_T|$ diffracted by a gap of width $2a$ in a break-
water: non-dimensional wavenumber $ka = \pi/2$, angle of wave
incidence $\beta = \pi/4$.

numerical scheme below. Graphical results for the diffraction coefficient
are presented in §5.1.2 and §7.2.

The representations (4.122)–(4.123) may be discretized in the manner
of (4.171) and used to compute the wave field throughout the fluid domain. A sample computation is given in Figure 4.3 where the modulus
of the total potential ϕ_T is illustrated; in the water-wave problem this
is proportional to the amplitude of the free surface oscillations. In the
figure, darker regions indicate small amplitude oscillations and lighter
regions indicate large amplitude oscillations. In $y > 0$ the principle
feature is the reflection of the incident wave from the barrier to create

standing waves which are indicated by the alternate light and dark bands roughly parallel to the x-axis.

A second numerical technique, known as the Galerkin method, for the solution of the integral equation for the breakwater-gap problem is now presented. In terms of the operator notation introduced in equation (4.152) the integral equation is

$$(Hv_\beta)(x) = f_\beta(x), \quad x \in L. \tag{4.173}$$

An approximation to the solution of this equation is sought in the form of a finite series

$$v_\beta(x) = \sum_{n=0}^{N} a_n \chi_n(x), \tag{4.174}$$

where the expansion set $\{\chi_n; \ n = 0, 1, \dots, N\}$ is prescribed and the coefficients $\{a_n; \ n = 0, 1, \dots, N\}$ are to be found. It is assumed here that the expansion set is real valued. The series (4.174) is substituted into (4.173) and then the inner product taken with each member of the expansion set to obtain

$$\sum_{n=0}^{N} a_n \langle H\chi_n, \chi_m \rangle = \langle f_\beta, \chi_m \rangle \equiv g_m(\beta), \quad m = 0, 1, \dots, N, \tag{4.175}$$

which is a set of N simultaneous equations for $\{a_n; \ n = 0, 1, \dots, N\}$. With the coefficients determined in this way, the diffraction coefficient

$$G(\theta, \beta) = \langle v_\beta, f_{\pi-\theta} \rangle = \sum_{n=0}^{N} a_n g_n(\theta). \tag{4.176}$$

In view of the known singular behaviour in $v_\beta(x)$ at $x = \pm a$ (see equation 4.114), an appropriate choice for the expansion set is

$$\chi_n(x) = \frac{T_n(x/a)}{(a^2 - x^2)^{1/2}}, \quad n = 0, 1, \dots, N, \tag{4.177}$$

where T_n denotes a Chebyshev polynomial of degree n. With this choice

$$\langle H\chi_n, \chi_m \rangle = \frac{i}{2} \int_L \int_L H_0^{(1)}(k|x - t|) \frac{T_n(x/a)}{(a^2 - x^2)^{1/2}} \frac{T_m(t/a)}{(a^2 - t^2)^{1/2}} \, dx \, dt \tag{4.178}$$

which with the changes of variable (4.168) is

$$\langle H\chi_n, \chi_m \rangle = \frac{i}{2} \int_0^\pi \int_0^\pi H_0^{(1)}(ka|\cos u - \cos v|) \cos nu \cos mv \, du \, dv.$$

$$(4.179)$$

By exploitation of the form of the integrand, it may be shown that $\langle H\chi_n, \chi_m \rangle = 0$ if $m+n$ is odd. Further

$$g_n(\beta) = \int_0^\pi e^{-ika \cos u \cos \beta} \cos nu \, du = \pi(-i)^n J_n(ka \cos \beta) \qquad (4.180)$$

(Gradshteyn and Ryzhik 1980, equations 3.715.13 and 3.715.18).

The imaginary part of the double integral in (4.179) has a non-singular integrand and is straightforward to evaluate numerically. The real part requires care as there is a singular line in the integrand that causes some difficulties. It is possible to rewrite this integral in terms of single integrals of products of Bessel functions, but as this introduces integrals with infinite ranges and oscillatory integrands there is probably no advantage in doing this. Instead, the real part can be written

$$\mathrm{Re}\langle H\chi_n, \chi_m \rangle = -\frac{1}{2} \int_0^\pi \int_0^\pi \left\{ Y_0(ka|\cos u - \cos v|) \right.$$

$$\left. -\frac{2}{\pi} \ln(|\cos u - \cos v|) \right\} \cos nu \cos mv \, du \, dv - \frac{1}{\pi} L_{nm}, \qquad (4.181)$$

where

$$L_{nm} = \int_0^\pi \int_0^\pi \ln(|\cos u - \cos v|) \cos nu \cos mv \, du \, dv$$

$$= \begin{cases} -\pi^2 \ln 2, & m = n = 0, \\ -\pi^2/2m, & m = n \neq 0, \\ 0, & m \neq n. \end{cases} \qquad (4.182)$$

(The last results follow from properties of the Chebyshev polynomials which are given, for example, by Estrada and Kanwal 1989, §2.) The integral given explicitly in (4.181) now has a non-singular integrand throughout the range of integration and is amenable to numerical integration by standard techniques.

Numerical results obtained using the Galerkin method are given in Table 4.2. Similar comments on convergence to those made on the collocation method apply here. The same parameters have been chosen for

			N		
ka	5	10	15	20	25
π	3.9883	3.9966	3.9966	3.9966	3.9966
2π	3.6659	5.7155	5.7156	5.7156	5.7156
3π	3.4499	7.0447	7.1336	7.1336	7.1336
4π	3.0742	5.6521	8.4435	8.4468	8.4468

Table 4.2 The modulus of the diffraction coefficient $G(11\pi/12, \pi/12)$ for various wavenumbers ka and truncation level N.

Tables 4.1 and 4.2 so that direct comparison between the two methods can be made. The convergence of the Galerkin method is quite rapid with increasing N and in all cases the results have converged to many more decimal places than those shown here. A disadvantage in using the Galerkin method for the present problem is that particular care must be taken in the computation of the double integral in equation (4.179). By contrast, the collocation method is very simple to implement and there is no difficulty in computing to graphical accuracy for the range of ka considered.

Bibliographical notes

Much of the material in §§4.7.1–4.7.2 is described by Lamb (1932, §305), although the integral equations are not given explicitly. The derivation of the integral equations is also described by Gilbert and Brampton (1985) who solve the equations by the collocation method given in §4.7.4. Gilbert and Brampton also give tables of numerical results for both the breakwater-gap and insular breakwater problems; however, their numerical results for the gap problem seem to contain some errors. Other authors have made similar calculations, for example Hunt (1990).

The problems of diffraction by an aperture in a screen with both a Neumann condition (the breakwater-gap problem) and a Dirichlet condition, and the complementary problems with a finite length strip, have exact solutions in terms of Mathieu functions (Carr and Stelzriede 1952, Morse and Rubinstein 1938). However, these solutions are little used due to the difficulty of computing Mathieu functions.

A related problem that has been solved using integral equations is that of diffraction by a circular harbour with a gap (Burrows 1985).

The problems of scattering by a breakwater consisting of a periodic row of gaps in a wall have been solved for normal incidence by Dalrymple and Martin (1990) using matched eigenfunction expansions, and for oblique incidence by Williams and Crull (1993) using an integral equation approach.

Williams (1982) discusses embedding formulas in both breakwater problems, but the approach used here in §4.7.3 follows Porter and Chu (1986). A direct method for obtaining the embedding formula for the breakwater-gap problem that does not rely on the link with the insular breakwater problem is described by Porter and Stirling (1994). An extension of these ideas to an infinite breakwater containing an arbitrary (finite) number of gaps has been made recently by Biggs, Porter, and Stirling (2000).

The Galerkin approach to the numerical solution of the integral equation used in §4.7.4 is described by Porter and Stirling (1990, chapter 8). Similar methods have been applied to wave diffraction by a gap between two breakwaters that are not in line by Smallman and Porter (1985).

4.8 Babinet's principle

The results of §4.7.3 show that the diffraction problems solved in §§4.7.1 and 4.7.2 are related. This is a form of Babinet's principle which, in the context of acoustic or electromagnetic waves, asserts that the problem of diffraction by a plane screen S containing apertures L is related to a complementary diffraction problem in which L is a screen with apertures S, although the boundary conditions in the two problems will be different. The precise relation between the problems for the electromagnetic case is given in Baker and Copson (1950, §3.3) and a discussion covering both the acoustic and electromagnetic cases is given by Jones (1986, §9.3). MacCamy (1958a) showed how this principle could be extended to a number of different types of scattering problem.

Following MacCamy, we will show how the solution to diffraction problems of the type considered in §4.7.2, where the scatterer is a finite strip, can be determined in terms of solutions to problems of the type treated in §4.7.1, where waves are diffracted by a gap in a plane screen.

If the problem of §4.7.1 is decomposed into parts which are, respectively, symmetric and antisymmetric about $y = 0$ it is readily shown

that it is equivalent to a half-plane problem for a potential ψ, assumed
to be bounded throughout the fluid domain, of the following type:

$$(\nabla^2 + k^2)\psi = 0 \qquad \text{in} \quad y > 0, \tag{4.183}$$

$$\frac{\partial\psi}{\partial y} = 0 \qquad \text{on} \quad y = 0, |x| > a, \tag{4.184}$$

$$\psi = h(x) \qquad \text{on} \quad y = 0, |x| < a, \tag{4.185}$$

$$\lim_{kr\to\infty} r^{1/2}\left(\frac{\partial\psi}{\partial r} - ik\psi\right) = 0. \tag{4.186}$$

Similarly, the problem treated in §4.7.2 is determined by the solution to
a half-plane problem of the form

$$(\nabla^2 + k^2)\phi = 0 \qquad \text{in} \quad y > 0, \tag{4.187}$$

$$\phi = 0 \qquad \text{on} \quad y = 0, |x| > a, \tag{4.188}$$

$$\frac{\partial\phi}{\partial y} = g(x) \qquad \text{on} \quad y = 0, |x| < a, \tag{4.189}$$

$$\lim_{kr\to\infty} r^{1/2}\left(\frac{\partial\phi}{\partial r} - ik\phi\right) = 0. \tag{4.190}$$

For any given functions g and h, analytic on $|x| < a$, the solution to
(4.187)–(4.190), designated P2, can be related to the solution to (4.183)–
(4.186), designated P1, in the following way.

Let $h_0(x)$ be any particular solution of the differential equation

$$\frac{d^2h}{dx^2} + k^2h = -g, \qquad |x| < a. \tag{4.191}$$

Assume that P1 can be solved for $h(x) = h_0(x)$, $h(x) = \exp(-ikx)$ and
$h(x) = \exp(ikx)$, and denote the respective solutions by ψ_0, ψ_1, and ψ_2.
Set

$$\phi = \frac{\partial\psi_0}{\partial y} + A\frac{\partial\psi_1}{\partial y} + B\frac{\partial\psi_2}{\partial y}, \tag{4.192}$$

where A and B are constants. By construction ϕ satisfies (4.188). Since
ψ_i, $i = 1, 2, 3$ all satisfy (4.183), it follows that

$$\frac{\partial^2\psi_i}{\partial y^2} = -\frac{\partial^2\psi_i}{\partial x^2} - k^2\psi_i^2, \qquad i = 1, 2, 3, \tag{4.193}$$

and this must hold on $y = 0$, $|x| < a$. Thus, on $y = 0$, $|x| < a$,

$$\frac{\partial \phi}{\partial y} = -\left(\frac{d^2}{dx^2} + k^2\right)(h_0(x) + A e^{-ikx} + B e^{ikx}) = g(x), \qquad (4.194)$$

where (4.191) has been used, and so ϕ also satisfies (4.189). The constants A and B can be determined by ensuring that ϕ has the appropriate behaviour near the points $(\pm a, 0)$.

As we have already remarked in §4.7.4, on $y = 0$

$$\frac{\partial \psi_i}{\partial y} \sim C_i^{\pm}(a \mp x)^{-1/2} \quad \text{as} \quad x \to \pm a, \qquad (4.195)$$

for some constants $C_i^{\pm}$. These numbers can be easily evaluated if the problem for ψ is solved using the Galerkin method. Thus if the normal derivative of ψ on the line $y = 0$, $|x| < a$, is approximated as

$$\frac{\partial \psi_i}{\partial y} = \sum_{n=0}^{N} a_n^{(i)} \chi_n(x), \qquad (4.196)$$

where χ_n is defined in (4.177), we can compute the coefficients $a_n^{(i)}$ from a system of equations of the same type as (4.175). It then follows that

$$\frac{\partial \psi_i}{\partial y} \sim [2a(a \mp x)]^{-1/2} \sum_{n=0}^{N} (\pm 1)^n a_n^{(i)} \quad \text{as} \quad x \to \pm a \qquad (4.197)$$

and so

$$C_i^{\pm} = (2a)^{-1/2} \sum_{n=0}^{N} (\pm 1)^n a_n^{(i)}. \qquad (4.198)$$

The solution to P2 must be regular at $(\pm a, 0)$; indeed it follows from (4.188) that $\phi(\pm a, 0) = 0$. But from (4.192) and (4.195) it is clear that

$$\phi \sim \left[C_0^{\pm} + A C_1^{\pm} + B C_2^{\pm}\right](a \mp x)^{-1/2} \quad \text{as} \quad x \to \pm a. \qquad (4.199)$$

Hence we must have

$$C_0^{\pm} + A C_1^{\pm} + B C_2^{\pm} = 0 \qquad (4.200)$$

and these two equations determine A and B.

For the specific problems solved in §§4.7.1 and 4.7.2 we have, from (4.116) and (4.124),

$$h(x) = -\,\mathrm{e}^{-\mathrm{i}kx\cos\beta} \quad \text{and} \quad g(x) = \mathrm{i}k\sin\beta\,\mathrm{e}^{-\mathrm{i}kx\cos\beta}, \qquad (4.201)$$

so we can take

$$h_0(x) = -\frac{\mathrm{i}}{k\sin\beta}\,\mathrm{e}^{-\mathrm{i}kx\cos\beta} \qquad (4.202)$$

and it follows that

$$\psi_0 = \frac{\mathrm{i}}{k\sin\beta}\varphi_\beta, \qquad (4.203)$$

where $\varphi_\beta(x,y)$ is the solution to the breakwater-gap problem (in $y > 0$) for angle of incidence β. Similarly ψ_1 is proportional to φ_0 and ψ_2 is proportional to φ_π. Hence for $\beta \neq 0$ or π we can write

$$\phi = \frac{\mathrm{i}}{k\sin\beta}\left(\frac{\partial\varphi_\beta}{\partial y} + \tilde{A}\frac{\partial\varphi_0}{\partial y} + \tilde{B}\frac{\partial\varphi_\pi}{\partial y}\right) \qquad (4.204)$$

and the constants $\tilde{A}$ and $\tilde{B}$ can be determined as the solutions to the equations

$$\tilde{C}_\beta^\pm + \tilde{A}\tilde{C}_0^\pm + \tilde{B}\tilde{C}_\pi^\pm = 0 \qquad (4.205)$$

in an obvious notation. This shows that the solution to the insular breakwater problem with angle of incidence β can be expressed (everywhere in the fluid domain) in terms of the solutions to the breakwater-gap problem with angles of incidence β, 0, and π. Of course, symmetry implies that $\varphi_\pi(x,y) = \varphi_0(-x,y)$ and $\tilde{C}_\pi^\pm = \tilde{C}_0^\mp$.

Bibliographical notes

Another example where a diffraction problem with a finite scatterer can be expressed in terms of the solution to a 'gap' problem is the infinite depth water-wave problem where the scattering structure is a finite dock (the equivalent problem in finite depth was examined using matched eigenfunction expansions in §2.5.3 and is treated using residue calculus theory in §5.2.2 below). For normal incidence, Rubin (1954) showed how the dock problem could be expressed in terms of the solution to a (non-physical) boundary-value problem in which the free surface is replaced by a rigid lid and the dock itself by a strip with a mixed boundary condition corresponding to (1.13) with negative K. One advantage of this approach is that the new problem, unlike the original one, is amenable to variational arguments and Rubin showed how the standard integral equation could be transformed into one with a much simpler kernel (though a more complicated right-hand side) and was able to use this new formulation to prove the existence and uniqueness of a solution. The transformed integral equation was used in Sparenberg (1957) and MacCamy (1961) to produce numerical solutions. For the case of oblique waves, the transformation can still be made (see MacCamy 1958a), but results were computed in Garrison (1969) from the more complicated integral equation.

Chapter 5

The Wiener-Hopf and related techniques

In some wave/structure interaction problems where the structure is semi-infinite in extent, it turns out that it is possible to obtain the solution explicitly in terms of an integral. This expression can be used as a basis for numerical calculations and it also allows certain information about the solution, such as the far-field form, to be obtained exactly.

The most commonly used method in diffraction theory for solving such problems is the Wiener-Hopf technique and this will be outlined in §5.1. In §5.1.1 the procedure will be illustrated by solving the classic Sommerfeld problem, namely the diffraction of waves by a rigid half-plane, and in §5.1.2 it will be shown how the solution to this problem can be used to obtain high-frequency approximations to the breakwater-gap and the insular breakwater problem given in §4.7. The problem of water-wave scattering by a submerged horizontal plate in finite depth is solved in §5.1.3 to illustrate the use of the Wiener-Hopf technique on problems involving the free-surface boundary condition (1.13).

An alternative method, which is perhaps conceptually simpler though not so widely applicable, is the so-called residue calculus theory and this is described in §5.2. Both the Wiener-Hopf technique and residue calculus theory are fairly technical procedures and some necessary pre-liminaries to the latter approach are discussed in §5.2.1. The first example of the application of the method is to the scattering of oblique waves by a rigid dock lying in the free surface. When the dock is semi-infinite in extent the problem can be solved exactly, but one of the advantages of the residue calculus theory is that it is easily modified to provide extremely efficient numerical methods for the solution of problems where the scatterer is finite. Thus in §5.2.2 it is shown how the problem that was formulated using matched eigenfunction expansions in

§2.5.3 can be transformed into one which is better suited to numerical calculations. This example also shows that the solution to semi-infinite problems can be used to provide good approximations to the equivalent problems involving scatterers of finite length and in §5.2.3 this idea is used to derive accurate approximations to the scattering of an incident wave by a comb-like grating.

5.1   The Wiener-Hopf technique

The Wiener-Hopf technique is a very powerful method which enables certain linear partial differential equations subject to boundary conditions on semi-infinite domains to be solved explicitly. It can also be used to provide approximate solutions to more realistic problems. The technique was initially developed to solve special types of integral equations and subsequently it was appreciated that certain problems in diffraction theory could be formulated as integral equations which were amenable to solution via the Wiener-Hopf procedure. The method as it is described below is based on the simplified procedure of Jones (1952), which removes the need to formulate the integral equation when solving partial differential equations.

Crucial to the method are the analyticity properties of complex Fourier transforms. Suppose first that s is real. The usual Fourier transform F of a function f and its inverse can be defined through the equations

$$F(s) = \int_{-\infty}^{\infty} f(x)\,\mathrm{e}^{\mathrm{i}sx}\,\mathrm{d}x, \tag{5.1}$$

$$f(x) = \frac{1}{2\pi} \int_{-\infty}^{\infty} F(s)\,\mathrm{e}^{-\mathrm{i}sx}\,\mathrm{d}s. \tag{5.2}$$

Suppose now that $s = \sigma + \mathrm{i}\tau$ is a complex variable. Under suitable conditions on f (see, for example, Noble 1958, p. 11) we can extend these relations and define the so-called generalized Fourier transform.

We begin by defining half-range transforms as follows. If $f(x) = 0$ for $x < 0$ and if $|f(x)| < A\exp(\tau_1 x)$ as $x \to \infty$, for some constants A (positive) and τ_1, then

$$F_+(s) = \int_0^{\infty} f(x)\,\mathrm{e}^{\mathrm{i}sx}\,\mathrm{d}x \tag{5.3}$$

is an analytic function of s in the region $\tau > \tau_1$ of the complex s-plane and

$$f(x) = \frac{1}{2\pi} \int_C F_+(s) \, e^{-isx} \, ds, \qquad (5.4)$$

where C is a path within the region of analyticity of $F_+(s)$ on which σ varies from $-\infty$ to ∞. Furthermore, if $f(x) \sim x^\lambda$ as $x \to 0^+$ ($\lambda > -1$), then

$$F_+(s) \sim \frac{\Gamma(\lambda+1)}{(-is)^{\lambda+1}} \qquad (5.5)$$

as $|s| \to \infty$ with $\tau > \tau_1$.

Similarly, if $f(x) = 0$ for $x > 0$ and if $|f(x)| < B \exp(\tau_2 x)$ as $x \to -\infty$, then

$$F_-(s) = \int_{-\infty}^0 f(x) \, e^{isx} \, dx \qquad (5.6)$$

is an analytic function of s in the region $\tau < \tau_2$ and

$$f(x) = \frac{1}{2\pi} \int_C F_-(s) \, e^{-isx} \, ds, \qquad (5.7)$$

where C is a path within the region of analyticity of $F_-(s)$ on which σ varies from $-\infty$ to ∞. If $f(x) \sim (-x)^\lambda$ as $x \to 0^-$ ($\lambda > -1$), then

$$F_-(s) \sim \frac{\Gamma(\lambda+1)}{(is)^{\lambda+1}} \qquad (5.8)$$

as $|s| \to \infty$ with $\tau < \tau_2$.

These results for half-range transforms can be combined. If

$$|f(x)| < \begin{cases} A \, e^{\tau_1 x} & \text{as} \quad x \to \infty, \\ B \, e^{\tau_2 x} & \text{as} \quad x \to -\infty, \end{cases} \qquad (5.9)$$

with $\tau_1 < \tau_2$, then the full-range transform defined by (5.1) is an analytic function of s in the strip $\tau \in (\tau_1, \tau_2)$ and the inversion formula is given by (5.2) with the path of integration lying within the strip of analyticity.

In a typical Wiener-Hopf solution procedure one begins by applying Fourier transforms to the underlying partial differential equation with respect to s to reduce the problem to an equation of the form

$$A(s)\Psi_+(s) + B(s)\Psi_-(s) + C(s) = 0, \qquad \tau \in (\tau_1, \tau_2), \qquad (5.10)$$

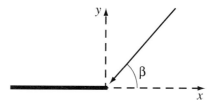

FIGURE 5.1
Diffraction by a half-plane.

in which A, B, and C are given functions whilst $\Psi_+(s)$ and $\Psi_-(s)$ are unknown functions with the properties that $\Psi_+(s)$ is analytic in the half-plane $\tau > \tau_1$ and $\Psi_-(s)$ is analytic in the half-plane $\tau < \tau_2$. The aim is then, by a suitable decomposition of the functions A, B, and C, to convert this to an equation of the form

$$K_+(s)\Psi_+(s) + C_+(s) = K_-(s)\Psi_-(s) + C_-(s), \quad \tau \in (\tau_1, \tau_2), \quad (5.11)$$

where all the 'plus' functions are analytic for $\tau > \tau_1$ and all the 'minus' functions are analytic for $\tau < \tau_2$. It is this decomposition which represents the crucial step in the process and which characterizes the difficulty of a given problem.

Although (5.11) is a single equation for the two unknowns $\Psi_\pm(s)$, it can be used to solve for both of these functions simultaneously. Each side of (5.11) defines the analytic continuation $J(s)$ of the other into the whole complex plane and knowledge of the behaviour of the functions in (5.11) as $|s| \to \infty$ in the strip of analyticity can then be used to determine $J(s)$ (and hence $\Psi_\pm(s)$).

Some of the technicalities associated with the Wiener-Hopf technique will be illustrated through the examples which follow. For more details the reader is referred to Noble (1958) and Crighton et al. (1992).

5.1.1   The Sommerfeld problem

Consider the diffraction of a plane wave by a semi-infinite rigid vertical barrier extending throughout water of constant depth h so that the bed is at $z = -h$. The incident wave makes an angle β with the x-axis and the barrier occupies $y = 0$, $x \in (-\infty, 0]$ as shown in Figure 5.1, which depicts a plan view of the geometry.

The depth variation for the problem can be factored out by writing

$$\Phi_T(x, y, z, t) = \text{Re}\left\{\phi_T(x, y) \cosh k(z + h) e^{-i\omega t}\right\}, \qquad (5.12)$$

where k is the solution of the dispersion relation (2.9) and then we can extract the incident wave field by setting

$$\phi_T(x, y) = \phi(x, y) + e^{-ikx \cos \beta - iky \sin \beta}. \qquad (5.13)$$

The boundary-value problem for ϕ is then

$$\nabla^2 \phi + k^2 \phi = 0 \qquad \text{in the fluid,} \qquad (5.14)$$

$$\frac{\partial \phi}{\partial y} = ik \sin \beta \, e^{-ikx \cos \beta} \qquad \text{on } x < 0, y = 0 \qquad (5.15)$$

and we note that $\partial \phi / \partial y$ is continuous on $y = 0$ for both $x > 0$ and $x < 0$ whereas ϕ is only continuous on $y = 0$ for $x > 0$. The appropriate edge condition is that

$$|\nabla \phi| = O(r^{-1/2}) \qquad \text{as} \qquad r = (x^2 + y^2)^{1/2} \to 0 \qquad (5.16)$$

and we must impose a radiation condition of the form (1.29) which requires the diffracted wave field to be outgoing. For mathematical convenience we treat k as a complex variable and write $k = k_1 + i\epsilon$, $\epsilon > 0$ (the limit $\epsilon \to 0$ will be taken at the end of the calculation) and then the radiation condition shows that (see Noble 1958, §2.2)

$$\phi = O\left(\exp[\epsilon x \cos \beta]\right) \qquad \text{as} \qquad x \to -\infty, \qquad (5.17)$$

$$\phi = O\left(x^{-1/2} \exp[-\epsilon x]\right) \qquad \text{as} \qquad x \to \infty. \qquad (5.18)$$

Next we introduce another complex variable $s = \sigma + i\tau$ and take the complex Fourier transform of (5.14) so that

$$\Psi''(s, y) - \gamma^2 \Psi(s, y) = 0, \qquad \gamma = (s^2 - k^2)^{1/2}, \qquad (5.19)$$

where

$$\Psi(s, y) = \int_{-\infty}^{\infty} \phi(x, y) e^{isx} \, dx \qquad (5.20)$$

and a prime indicates differentiation with respect to y. The branch of γ is chosen so that when $s = 0$, $\gamma = -ik$ and $\gamma \sim |s|$ as $s \to \pm\infty$

along the real axis. The full-range Fourier transform Ψ is decomposed as $\Psi = \Psi_+ + \Psi_-$, where

$$\Psi_+(s, y) = \int_0^\infty \phi(x, y) \, e^{isx} \, dx, \tag{5.21}$$

$$\Psi_-(s, y) = \int_{-\infty}^0 \phi(x, y) \, e^{isx} \, dx \tag{5.22}$$

and (5.17), (5.18) show that Ψ_+ is an analytic function of s provided $\tau > -\epsilon$ whereas Ψ_- is an analytic function of s provided $\tau < \epsilon \cos \beta$.

Since the real part of γ is always positive in the strip $\tau \in (-\epsilon, \epsilon)$ the solution of (5.19) subject to the requirement of boundedness in y and the continuity of $\partial \phi / \partial y$ (and hence of $\partial \Psi / \partial y$) on $y = 0$ is

$$\Psi(s, y) = \begin{cases} A(s) \, e^{-\gamma y} & y \geq 0, \\ -A(s) \, e^{\gamma y} & y < 0, \end{cases} \tag{5.23}$$

where $A(s)$ is as yet undetermined. The continuity conditions on ϕ show that the functions $\Psi_+(s, 0)$, $\Psi_+'(s, 0)$, $\Psi_-'(s, 0)$ are well defined and that, from (5.23),

$$\Psi_+(s, 0) + \Psi_-(s, 0^+) = \Psi(s, 0^+) = A(s), \tag{5.24}$$

$$\Psi_+(s, 0) + \Psi_-(s, 0^-) = \Psi(s, 0^-) = -A(s), \tag{5.25}$$

$$\Psi_+'(s, 0) + \Psi_-'(s, 0) = \Psi'(s, 0) = -\gamma A(s), \tag{5.26}$$

in which $\Psi(s, 0^\pm) = \lim_{y \to 0\pm} \Psi(s, y)$ etc. The only function in these equations for which the region of analyticity is unknown is $A(s)$ so we choose to eliminate it. From (5.15),

$$\Psi_-'(s, 0) = \frac{k \sin \beta}{(s - k \cos \beta)} \tag{5.27}$$

and so (5.24)–(5.26) reduce to

$$\Psi_+(s, 0) = -S_-(s), \tag{5.28}$$

$$\Psi_+'(s, 0) + \frac{k \sin \beta}{(s - k \cos \beta)} = -\gamma D_-(s), \tag{5.29}$$

where

$$S_-(s) = \frac{1}{2} \left[\Psi_-(s, 0^+) + \Psi_-(s, 0^-) \right], \tag{5.30}$$

$$D_-(s) = \frac{1}{2} \left[\Psi_-(s, 0^+) - \Psi_-(s, 0^-) \right]. \tag{5.31}$$

Both (5.28) and (5.29) hold in the strip $\tau \in (-\epsilon, \epsilon \cos \beta)$ and are of standard Wiener-Hopf type (5.10) but we actually only need to solve (5.29) since knowledge of $\Psi'_+(s,0)$ is sufficient to determine $A(s)$ from (5.26) and (5.27).

In order to transform (5.29) into an equation of the form (5.11) we proceed as follows. First

$$\gamma = (s+k)^{1/2}(s-k)^{1/2} \tag{5.32}$$

in which the first factor is analytic for $\tau > -\epsilon$ and the second is analytic for $\tau < \epsilon$ and secondly

$$\frac{k \sin \beta}{(s - k \cos \beta)(s+k)^{1/2}} = H_+(s) + H_-(s), \tag{5.33}$$

where

$$H_+(s) = \frac{k \sin \beta}{(s - k \cos \beta)} \left(\frac{1}{(s+k)^{1/2}} - \frac{1}{(k + k \cos \beta)^{1/2}} \right) \tag{5.34}$$

is analytic for $\tau > -\epsilon$ and

$$H_-(s) = \frac{k \sin \beta}{(s - k \cos \beta)(k + k \cos \beta)^{1/2}} \tag{5.35}$$

is analytic for $\tau < \epsilon \cos \beta$. Equation (5.29) can then be rearranged to give

$$\frac{\Psi'_+(s,0)}{(s+k)^{1/2}} + H_+(s) = -H_-(s) - (s-k)^{1/2} D_-(s) = J(s), \tag{5.36}$$

where $J(s)$ is analytic in the whole s-plane. To determine J we need to know the behaviour of the functions in (5.36) as $|s| \to \infty$ in the strip $\tau \in (-\epsilon, \epsilon \cos \beta)$.

From (5.5), (5.8), and (5.16),

$$\Psi'_+(s,0) = O(s^{-1/2}) \quad \text{as} \quad |s| \to \infty, \tau > -\epsilon, \tag{5.37}$$

$$\Psi_-(s,0^\pm) = O(s^{-1}) \quad \text{as} \quad |s| \to \infty, \tau < \epsilon \cos \beta \tag{5.38}$$

and by construction $H_\pm(s) = O(s^{-1})$ as $|s| \to \infty$ in the appropriate region of the s-plane. It follows that $J(s)$ is analytic in the whole s-plane and tends to zero as $s \to \infty$ in any direction. Hence from Liouville's theorem $J(s)$ is identically zero (see, for example, Noble 1958, p. 6).

From (5.26), (5.27), (5.34), and (5.36), $A(s)$ can be determined and then (5.23) and the Fourier inversion theorem show that

$$\phi(x, y) = -\frac{\operatorname{sgn} y}{2\pi}(k - k \cos \beta)^{1/2} \int_{-\infty+ia}^{\infty+ia} \frac{\exp(-\mathrm{i}sx - \gamma \operatorname{sgn} y)}{(s - k)^{1/2}(s - k \cos \beta)} \, \mathrm{d}s,$$
(5.39)

where $a \in (-\epsilon, \epsilon \cos \beta)$. This solution is discussed in detail by Noble, (1958, §1.6 and §2.6) and he shows that it can be expressed in terms of Fresnel integrals. With $x = r \cos \theta$, $y = r \sin \theta$, $\theta \in (-\pi, \pi)$,

$$\phi(r, \theta) = \frac{\mathrm{e}^{-\mathrm{i}\pi/4}}{\pi^{1/2}} \left\{ \mathrm{e}^{-\mathrm{i}kr \cos(\theta+\beta)} F\left[(2kr)^{1/2} \cos \frac{\theta + \beta}{2}\right] \right.$$
$$\left. - \mathrm{e}^{-\mathrm{i}kr \cos(\theta-\beta)} F\left[(2kr)^{1/2} \cos \frac{\theta - \beta}{2}\right] \right\}, \quad (5.40)$$

where

$$F(v) = \int_v^\infty \exp(\mathrm{i}u^2) \, \mathrm{d}u.$$
(5.41)

5.1.2   High-frequency approximations

In §4.7 numerical methods were described for the solutions of the problems of diffraction of waves by a breakwater with a gap and by an insular breakwater. High-frequency approximations to the solution of both of these problems can be obtained by suitable combinations of the exact solution to the problem of diffraction of waves by a rigid half-plane given above in equation (5.40). For example, in the breakwater-gap problem, if the frequency of the waves is sufficiently high there will be little interaction between the two ends of the gap and an approximation is obtained by combining two solutions for a half-plane in a suitable way. For a gap or insular breakwater of width $2a$ and an incident wave of wavenumber k, high frequency corresponds to the limit $ka \to \infty$. This limit could equally well be interpreted as corresponding to an increasingly large gap or breakwater in waves of fixed wavelength.

For a half-plane occupying $\{y = 0, x \leq 0\}$ and subject to an incident wave $\phi_{\mathrm{I}}(r, \theta; \beta)$ given by equation (4.111), the total potential may be written

$$\phi_{\mathrm{T}}(r, \theta; \beta) = \phi_{\mathrm{I}}(r, \theta; \beta) + \phi(r, \theta; \beta),$$
(5.42)

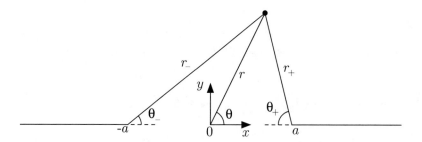

FIGURE 5.2
Coordinate systems for the breakwater-gap problem.

where the scattered field ϕ is given by equation (5.40). For real v, the Fresnel integral has the asymptotic forms

$$F(v) \sim \frac{i\,e^{iv^2}}{2v} \quad \text{and} \quad F(-v) \sim \pi^{1/2}\,e^{i\pi/4} - \frac{i\,e^{iv^2}}{2v} \quad \text{as} \quad v \to \infty \quad (5.43)$$

(Jones 1994, §8.13) and therefore

$$\phi(r,\theta;\beta) = \begin{cases} \phi_S(r,\theta;\beta), & \theta \in (-\pi, \pi - \theta_0), \\ \phi_S(r,\theta;\beta) + e^{-ikr\cos(\theta+\beta)}, & \theta \in (\pi - \theta_0, \pi), \end{cases} \quad (5.44)$$

where as $kr \to \infty$,

$$\phi_S(r,\theta;\beta) \sim \left(\frac{2}{\pi kr}\right)^{1/2} \frac{e^{i(kr+\pi/4)} \sin\frac{1}{2}\theta \sin\frac{1}{2}\beta}{\cos\theta + \cos\beta}. \quad (5.45)$$

As previously in §4.7.1, for the breakwater-gap problem the origin of coordinates is chosen in the centre of the gap. It is convenient to introduce further polar coordinate systems $(r_\pm, \theta_\pm)$ with origins at $(x, y) = (\pm a, 0)$ as shown in Figure 5.2. Now

$$x = \pm a \mp r_\pm \cos\theta_\pm \quad \text{and} \quad y = r_\pm \sin\theta_\pm, \quad (5.46)$$

so that

$$\phi_I(r,\theta;\beta) = e^{ika\cos\beta}\,\phi_I(r_-,\theta_-;\beta) \quad (5.47)$$

$$= e^{-ika\cos\beta}\,\phi_I(r_+,\theta_+;\pi-\beta) \quad (5.48)$$

and hence, under the assumption that the wave fields diffracted from each end of the gap do not interact, an approximation to the solution in

$y > 0$ is

$$\phi_{\mathrm{T}}(r, \theta; \beta) \approx \phi_{\mathrm{I}}(r, \theta; \beta) + \mathrm{e}^{\mathrm{i}ka\cos\beta}\, \phi(r_-, \theta_-; \beta)$$
$$+ \mathrm{e}^{-\mathrm{i}ka\cos\beta}\, \phi(r_+, \theta_+; \pi - \beta), \quad (5.49)$$

where ϕ is given by (5.40). As $r/a \to \infty$ for $\theta \in (0, \pi)$

$$r_{\pm} = r\left[1 \mp \frac{a}{r}\cos\theta + O\left(\frac{a^2}{r^2}\right)\right], \quad \theta_- \sim \theta \quad \text{and} \quad \theta_+ \sim \pi - \theta, \quad (5.50)$$

so that from equations (5.44)–(5.45),

$$\phi_{\mathrm{T}}(r, \theta; \beta) = \mathrm{e}^{-\mathrm{i}kr\cos(\theta - \beta)} + \mathrm{e}^{-\mathrm{i}kr\cos(\theta + \beta)} + \phi_+(r, \theta; \beta) \quad (5.51)$$

with

$$\phi_+(r, \theta; \beta) \sim \left(\frac{2}{\pi kr}\right)^{1/2} \frac{\mathrm{e}^{\mathrm{i}(kr + \pi/4)}}{\cos\theta + \cos\beta} \left\{ \mathrm{e}^{\mathrm{i}ka(\cos\theta + \cos\beta)} \sin\tfrac{1}{2}\theta \sin\tfrac{1}{2}\beta \right.$$
$$\left. - \mathrm{e}^{-\mathrm{i}ka(\cos\theta + \cos\beta)} \cos\tfrac{1}{2}\theta \cos\tfrac{1}{2}\beta \right\} \quad \text{as} \quad kr \to \infty. \quad (5.52)$$

The last expression is undefined when $\theta = \pi - \beta$; however, an application of L'Hôpital's rule gives

$$\phi_+(r, \pi - \beta; \beta) \sim \frac{\mathrm{e}^{\mathrm{i}(kr + \pi/4)}}{(2\pi kr)^{1/2}} \left\{ 2\mathrm{i}ka\sin\beta - \frac{1}{\sin\beta} \right\} \quad \text{as} \quad kr \to \infty$$
$$(5.53)$$

which demonstrates that the approximation breaks down when $\beta = 0$ or π. The diffraction coefficient $G(\theta, \beta)$ follows immediately from (4.129).

For the insular breakwater, a similar calculation leads to the approximation

$$\phi_{\mathrm{T}}(r, \theta; \beta) = \mathrm{e}^{-\mathrm{i}kr\cos(\theta - \beta)} + \phi(r, \theta; \beta), \quad \theta \in (0, \pi), \quad (5.54)$$

where

$$\phi(r, \theta; \beta) \sim \left(\frac{2}{\pi kr}\right)^{1/2} \frac{\mathrm{e}^{\mathrm{i}(kr - 3\pi/4)}}{\cos\theta + \cos\beta} \left\{ \mathrm{e}^{\mathrm{i}ka(\cos\theta + \cos\beta)} \cos\tfrac{1}{2}\theta \cos\tfrac{1}{2}\beta \right.$$
$$\left. - \mathrm{e}^{-\mathrm{i}ka(\cos\theta + \cos\beta)} \sin\tfrac{1}{2}\theta \sin\tfrac{1}{2}\beta \right\} \quad \text{as} \quad kr \to \infty. \quad (5.55)$$

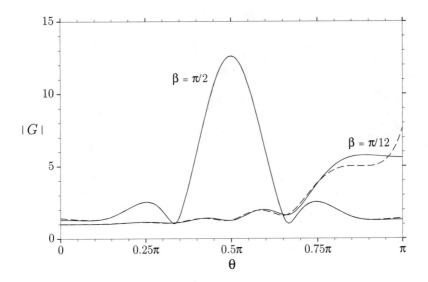

FIGURE 5.3
The modulus of the diffraction coefficient $G(\theta, \beta)$ for a break-water with a gap with $ka = 2\pi$. Accurate numerical results are shown with a solid line and the high-frequency approximation with a dashed line.

The last expression is undefined when $\theta = \pi - \beta$; however, an application of L'Hôpital's rule gives

$$\phi(r, \pi - \beta; \beta) \sim \frac{e^{i(kr-3\pi/4)}}{(2\pi kr)^{1/2}} \left\{ 2ika \sin \beta + \frac{1}{\sin \beta} \right\} \quad \text{as} \quad kr \to \infty$$

$$(5.56)$$

and as before the approximation breaks down when $\beta = 0$ or π. The diffraction coefficient $F(\theta, \beta)$ follows from (4.147).

The above high-frequency approximations are compared in Figures 5.3 and 5.4 with numerical solutions of the integral equations obtained in §4.7 for a moderately high value $ka = 2\pi$. For both the breakwater gap and insular breakwater problems the high-frequency approximation performs reasonably well for normal incidence $\beta = \pi/2$. The accuracy of the approximations deteriorates as β is reduced towards grazing incidence. However, the comparisons for the insular breakwater are generally poorer than for the gap problem. First of all, it is apparent from equation (5.55) that the high-frequency approximation to the insular

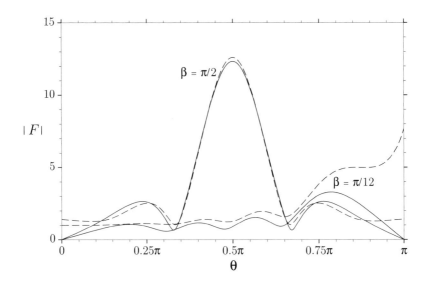

FIGURE 5.4
The modulus of the diffraction coefficient $F(\theta, \beta)$ for an insular breakwater with $ka = 2\pi$. Accurate numerical results are shown with a solid line and the high-frequency approximation with a dashed line.

breakwater problem does not satisfy (4.138) and hence there are discrepancies near $\theta = 0, \pi$. Secondly, the neglected interaction terms in the high-frequency expansion of the solution are larger in the insular breakwater problem. Kobayashi (1991, §2) gives a thorough analysis of this problem and shows the interaction terms to be of order $(ka)^{-1/2}$ as $ka \to \infty$. Abrahams (1982) has examined the gap problem (for normal incidence) and shows the interaction terms to be of order $(ka)^{-3/2}$ as $ka \to \infty$.

Bibliographical notes

The high-frequency approximation used here is formulated by Jones (1986, §9.25) and, in the context of breakwaters, by Penney and Price (1952) although the latter authors consider the case of normal incidence only. This approximation assumes there is no interaction between the ends of the gap or breakwater. Noble (1958, §5.6) describes a method of obtaining high-frequency approximations that includes an interaction term for a variety of diffraction problems. The method is applicable to the problems discussed here, but Noble works out the details only for a finite-length strip with a boundary condition corresponding to zero potential on the strip. This last problem had been examined previously by Karp and Russek (1956) using a technique based on physical arguments. High-frequency approximations including interaction terms are given for the rigid insular breakwater by Kobayashi (1991, §2) and for the rigid breakwater with a gap, under normal incidence, by Abrahams (1982).

5.1.3 A submerged horizontal plate

The Wiener-Hopf technique can also be used to solve problems involving the free-surface boundary condition (1.13). Consider the case of a fixed thin rigid plate placed along $z = -d$, $x \leq 0$, $y \in (-\infty, \infty)$ in water of depth h ($h > d$), with an oblique wave incident from $x = -\infty$ above the plate. A factor $\exp(\mathrm{i}\ell y)$ can be removed as described in §2.3.2 and we are left with the problem of solving the two-dimensional modified Helmholtz equation

$$(\nabla^2 - \ell^2)\varphi(x, z) = 0 \quad \text{in the fluid,} \tag{5.57}$$

together with the boundary conditions

$$\frac{\partial \varphi}{\partial z} = K\varphi \quad \text{on} \quad z = 0, \tag{5.58}$$

$$\frac{\partial \varphi}{\partial z} = 0 \quad \text{on} \quad z = -h, \tag{5.59}$$

$$\frac{\partial \varphi}{\partial z} = 0 \quad \text{on} \quad z = -d, \, x < 0 \tag{5.60}$$

and the edge condition

$$|\nabla \varphi| = O(r^{-1/2}) \quad \text{as} \quad r = \left[x^2 + (z - d)^2\right]^{1/2} \to 0. \tag{5.61}$$

In the region $x > 0$, propagating waves take the form

$$\left(A \, \mathrm{e}^{\mathrm{i}\alpha x} + B \, \mathrm{e}^{-\mathrm{i}\alpha x}\right) \psi_0(z),$$

where $\alpha = (k^2 - \ell^2)^{1/2}$, k is the positive solution to the dispersion relation $k \tanh kh = K$, and we clearly require $k > \ell$ for such waves to exist. In the region above the plate the situation is very similar except that the water depth is now d, so we define κ to be the positive solution to the dispersion relation $\kappa \tanh \kappa d = K$ and then propagating waves in $x < 0$, $z \in (-d, 0)$, take the form

$$\left(\tilde{A} \, \mathrm{e}^{\mathrm{i}\mu x} + \tilde{B} \, \mathrm{e}^{-\mathrm{i}\mu x}\right) \tilde{\psi}_0(z),$$

where $\mu = (\kappa^2 - \ell^2)^{1/2}$ and $\tilde{\psi}_0(z)$ is defined as in (2.11) but with h replaced by d. It follows that we require $\kappa > \ell$ for propagating waves to be possible in the region above the plate, but if $k < \ell < \kappa$ no waves can propagate into the region $x > 0$ and so there will be total reflection. For simplicity we will restrict attention to this case. The reflection coefficient R, which is defined in terms of the far-field behaviour through

$$\varphi \sim \tilde{\psi}_0(z) \left(e^{i\mu x} + R e^{-i\mu x} \right) \quad \text{as} \quad x \to -\infty, \ -d < z < 0, \qquad (5.62)$$

thus satisfies $|R| = 1$. As $x \to \infty$ and as $x \to -\infty$ below the plate we require φ to tend to zero. For mathematical convenience we treat μ as a complex variable and write $\mu = \mu_1 + i\epsilon$, $\epsilon > 0$, so as to ensure the convergence of certain integrals. The original problem can then be recovered by letting $\epsilon \to 0$ at the end.

Let the velocity potential be given by

$$\varphi(x, z) = \begin{cases} e^{i\mu x} \, \tilde{\psi}_0(z) + \varphi_1(x, z) & -d < z < 0, \\ \varphi_2(x, z) & -h < z < -d \end{cases} \qquad (5.63)$$

and define full- and half-range Fourier transforms by

$$\Phi_j(s, z) = \int_{-\infty}^{\infty} \varphi_j(s, z) \, e^{isx} \, dx, \quad j = 1, 2, \qquad (5.64)$$

$$= \int_{0}^{\infty} \varphi_j(s, z) \, e^{isx} \, dx + \int_{-\infty}^{0} \varphi_j(s, z) \, e^{isx} \, dx \qquad (5.65)$$

$$= \Phi_{j+}(s, z) + \Phi_{j-}(s, z). \qquad (5.66)$$

It follows from (5.62) and (5.63) that, provided ϵ is small enough,

$$\varphi_1 = O\left(\exp[-\epsilon|x|] \right) \quad \text{as} \quad |x| \to \infty \qquad (5.67)$$

and, in view of the nature of the eigenfunction expansion that can be constructed for φ, φ_2 also decays exponentially as $|x| \to \infty$. Hence $\Phi_{j+}(s, z)$ exists and is analytic in the half-plane $\tau > -\epsilon$ and $\Phi_{j-}(s, z)$ exists and is analytic in the half-plane $\tau < \epsilon$. Subscripts $+$ and $-$ will be used throughout this analysis to indicate functions which are analytic in the regions $\tau > -\epsilon$ and $\tau < \epsilon$, respectively.

If we take the Fourier transform of (5.57) and use a prime to denote differentiation with respect to z, we obtain

$$\Phi_j''(s, z) - \gamma^2 \Phi_j(s, z) = 0, \qquad j = 1, 2, \qquad (5.68)$$

where $\gamma(s) = (s^2 + \ell^2)^{1/2}$, suitably defined so as to be single valued. Unlike in the Sommerfeld problem considered previously, the precise definition of the branch of γ will not concern us here; this is because the function $K(s)$ which appears in (5.78) below is actually a function of γ^2. So as to satisfy the free-surface and bottom boundary conditions, let

$$\Phi_1(s, z) = A(s)\{\gamma \cosh \gamma z + K \sinh \gamma z\}, \quad -d < z < 0, \quad (5.69)$$
$$\Phi_2(s, z) = B(s) \cosh \gamma(z + h), \quad -h < z < -d, \quad (5.70)$$

where $A(s)$ and $B(s)$ are to be determined.

The boundary condition on the plate, the continuity of vertical velocity and the continuity of pressure across $z = -d$, $x > 0$ transform to

$$\Phi'_{1-}(s, -d) = \Phi'_{2-}(s, -d) = 0, \quad (5.71)$$
$$\Phi'_{1+}(s, -d) = \Phi'_{2+}(s, -d) = Q_+(s), \quad \text{say}, \quad (5.72)$$

and

$$\Phi_{1+}(s, -d) - \Phi_{2+}(s, -d) + \frac{i\tilde{\psi}_0(-d)}{\mu + s} = 0. \quad (5.73)$$

From (5.71) and (5.72) it follows that the problem is completely determined once $Q_+(s)$ is known since differentiating equations (5.69) and (5.70) with respect to z and then putting $z = -d$ shows that

$$A(s) = -Q_+(s)\gamma^{-1}(\gamma \sinh \gamma d - K \cosh \gamma d)^{-1}, \quad (5.74)$$
$$B(s) = Q_+(s)(\gamma \sinh \gamma c)^{-1}, \quad (5.75)$$

where $c = h - d$. Substituting these expressions into equations (5.69) and (5.70), subtracting and using (5.66) and (5.73) gives

$$D_-(s, -d) + \frac{i\tilde{\psi}_0(-d)}{\mu + s} = Q_+(s)K(s), \quad (5.76)$$

where

$$D_-(s, z) = \Phi_{2-}(s, z) - \Phi_{1-}(s, z) \quad (5.77)$$

and

$$K(s) = \frac{\gamma \sinh \gamma h - K \cosh \gamma h}{\gamma \sinh \gamma c \, (\gamma \sinh \gamma d - K \cosh \gamma d)}. \quad (5.78)$$

The next step in the Wiener-Hopf procedure is to factorize $K(s)$ into a product of two functions, one a 'plus' function analytic in $\tau > -\epsilon$ and the other a 'minus' function analytic in $\tau < \epsilon$. Thus we write

$$K(s) = K_+(s)K_-(s), \qquad (5.79)$$

where as yet the functions K_+ and K_- are undetermined.

Equation (5.76) can then be rearranged to give

$$\frac{1}{K_-(s)}\left(D_-(s,-d) + \frac{i\tilde{\psi}_0(-d)}{\mu+s}\right) - \frac{i\tilde{\psi}_0(-d)}{K_-(-\mu)(\mu+s)}$$
$$= Q_+(s)K_+(s) - \frac{i\tilde{\psi}_0(-d)}{K_-(-\mu)(\mu+s)}, \qquad (5.80)$$

which has the property that the left-hand side is analytic in the region $\tau < \epsilon$ whilst the right-hand side is analytic in the region $\tau > -\epsilon$ of the complex s-plane. It follows that (5.80) is valid in the strip $-\epsilon < \tau < \epsilon$ and that both sides can be analytically continued to give a function $J(s)$ which is analytic throughout the complex s-plane.

The required product factorization given by (5.79) is carried out as follows. The function $K(s)$ defined by (5.78) has zeros and poles but no branch points in the complex s-plane and we write $K(s)$ using an infinite product representation which exhibits its poles and zeros explicitly. Beginning with the factor $\gamma \sinh \gamma c$ in the denominator we note that the infinite product representation of $\sinh z$ is given in Abramowitz and Stegun (1965, eqn 4.5.68) and thus we can write

$$\gamma c \sinh \gamma c = \left\{ c(s+i\ell) \prod_{n=1}^{\infty} \left([1+\ell^2/\lambda_n^2]^{1/2} - is/\lambda_n\right) e^{is/\lambda_n} \right\}$$
$$\times \left\{ c(s-i\ell) \prod_{n=1}^{\infty} \left([1+\ell^2/\lambda_n^2]^{1/2} + is/\lambda_n\right) e^{-is/\lambda_n} \right\}, \qquad (5.81)$$

where $\lambda_n = n\pi/c$ and the function enclosed by the first pair of braces is non-zero and analytic in the region $\tau > -\epsilon$ and thus represents a 'plus' function whilst the function in the second pair of braces is a 'minus' function, non-zero and analytic for $\tau < \epsilon$. The exponential factors are included to ensure the absolute convergence of the two infinite products.

The function $f(z) = \Lambda \cosh z - z \sinh z$ can also be written as an infinite product. The zeros of this function are, as described in §2.1, $\pm z_n$, $n = 0, 1, 2, \ldots$, where z_0 is real, positive, and satisfies $z_0 \tanh z_0 = \Lambda$

and z_n, $n > 0$, are purely imaginary (with positive imaginary part) and satisfy $z_n \tan z_n + \Lambda = 0$. That these are the only zeros can be established by using the argument principle of complex variable theory; see for example McIver (1998). Thus the Weierstrass product expansion for $f(z)$ is

$$\Lambda \cosh z - z \sinh z = \Lambda \prod_{n=0}^{\infty} \left(1 - \frac{z^2}{z_n^2}\right). \tag{5.82}$$

It follows that

$$\gamma h \sinh \gamma h - K h \cosh \gamma h = K h \left(\frac{\gamma^2}{k^2} - 1\right) \prod_{n=1}^{\infty} \left(1 + \frac{\gamma^2}{k_n^2}\right), \tag{5.83}$$

where k_n are positive and satisfy $k_n \tan k_n h + K = 0$. The function $\gamma h \sinh \gamma h - K h \cosh \gamma h$ can thus be written as the product of a 'plus' function and a 'minus' function, defined by

$$(\gamma h \sinh \gamma h - K h \cosh \gamma h)_{\pm} = (Kh)^{1/2} \left(s/k \pm \mathrm{i} \left[\ell^2/k^2 - 1\right]^{1/2}\right)$$
$$\times \prod_{n=1}^{\infty} \left(\left[1 + \ell^2/k_n^2\right]^{1/2} \mp \mathrm{i} s/k_n\right) \mathrm{e}^{\pm \mathrm{i} sh/n\pi} \tag{5.84}$$

in an obvious notation. The behaviour of the numbers k_n for large n, given by (2.6) ensures that the infinite products are absolutely convergent. Similarly we have

$$(\gamma d \sinh \gamma d - K d \cosh \gamma d)_{\pm} = (Kd)^{1/2} \kappa^{-1} (s \pm \mu)$$
$$\times \prod_{n=1}^{\infty} \left(\left[1 + \ell^2/\kappa_n^2\right]^{1/2} \mp \mathrm{i} s/\kappa_n\right) \mathrm{e}^{\pm \mathrm{i} sd/n\pi}, \tag{5.85}$$

where the numbers κ_n are positive and satisfy $\kappa_n \tan \kappa_n d + K = 0$.

We can therefore write

$$K_{\pm}(s) = (cd/h)^{1/2} \frac{(\gamma h \sinh \gamma h - K h \cosh \gamma h)_{\pm} \, \mathrm{e}^{\pm \chi(s)}}{(\gamma c \sinh \gamma c)_{\pm} (\gamma d \sinh \gamma d - K d \cosh \gamma d)_{\pm}}, \tag{5.86}$$

where the exponential factor $\exp(\pm \chi(s))$ is to be chosen so that the functions $K_{\pm}(s)$ can be made to have algebraic (rather than exponential) growth in their respective half-planes as $|s| \to \infty$. The determination of $\chi(s)$ requires a careful analysis of the asymptotic forms of the various infinite products that are implicit in (5.86). Results from §5.2.1 below show that the correct choice for $\chi(s)$ is given by

$$\chi(s) = (is/\pi)(c\ln c + d\ln d - h\ln h) \tag{5.87}$$

and this choice implies that

$$K_\pm(s) = O(s^{-1/2}) \quad \text{as } |s| \to \infty. \tag{5.88}$$

It then follows from (5.5), (5.8), and (5.61) that both sides of equation (5.80) tend to zero as $|s| \to 0$ in the strip of analyticity and thus from Liouville's theorem (Noble 1958, p. 6) that $J(s) = 0$. Hence

$$Q_+(s) = -\frac{i\tilde{\psi}_0(-d)}{(\mu + s)K_+(s)K_+(\mu)}, \tag{5.89}$$

where the fact that $K_+(s) = -K_-(-s)$, which follows from (5.81) and (5.84)–(5.86), has been used.

Substituting this form for $Q_+(s)$ into equations (5.74) and (5.69) and then using the inversion formula for Fourier transforms results in

$$\varphi_1(x, z) = \frac{i\tilde{\psi}_0(-d)}{2\pi K_+(\mu)} \int_P \frac{(\gamma\cosh\gamma z + K\sinh\gamma z)\,e^{-isx}}{\gamma(\gamma\sinh\gamma d - K\cosh\gamma d)K_+(s)(\mu + s)}\,ds, \tag{5.90}$$

where P is a path from $\sigma = -\infty$ to ∞ in the strip $-\epsilon < \tau < \epsilon$ which passes above any singularities on the negative real axis and below any on the positive real axis. In the limit $\epsilon \to 0$ this path becomes the real axis itself with suitable indentations to avoid any singularities on the axis.

In order to calculate the reflection coefficient we must examine the form of $\varphi_1(x, z)$ for $x < 0$. Closing the contour in the upper half plane we pick up contributions from the poles corresponding to the zeros of $\gamma\sinh\gamma d - K\cosh\gamma d$ which lie on the positive real axis or have positive imaginary part, i.e. those at $s = \mu$ and $s = i(\kappa_n^2 + \ell^2)^{1/2}$, $n = 1, 2, 3, \ldots$. The contributions from all these poles are exponentially small as $x \to -\infty$ except that from $s = \mu$ (which corresponds to $\gamma = \kappa$) and thus we obtain

$$\varphi_1(x, z) \sim \tilde{\psi}_0(z)R\,e^{-i\mu x}, \tag{5.91}$$

where, making use of the fact that $\kappa\tanh\kappa d = K$,

$$\frac{1}{R} = -\left(1 + \frac{\sinh 2\kappa d}{2\kappa d}\right)2\mu^2 dK_+^2(\mu). \tag{5.92}$$

From (5.81) and (5.84)–(5.86) we have that

$$K_+(\mu) = e^{\chi(\mu)} \frac{c^{-1/2}\kappa}{2\mu(\mu + i\ell)} \left(\mu/k + i\left[\ell^2/k^2 - 1\right]^{1/2}\right)$$

$$\times \prod_{n=1}^{\infty} \frac{\left(\left[1 + \ell^2/k_n^2\right]^{1/2} - i\mu/k_n\right)}{\left(\left[1 + \ell^2/\kappa_n^2\right]^{1/2} - i\mu/\kappa_n\right)\left(\left[1 + \ell^2/\lambda_n^2\right]^{1/2} - i\mu/\lambda_n\right)} \quad (5.93)$$

from which, since χ is purely imaginary and $\mu^2 + \ell^2 = \kappa^2$,

$$|K_+(\mu)|^2 = \frac{\left(\kappa^2/k^2 - 1\right)}{4c\mu^2} \prod_{n=1}^{\infty} \frac{\left(1 + \kappa^2/k_n^2\right)}{\left(1 + \kappa^2/\kappa_n^2\right)\left(1 + \kappa^2/\lambda_n^2\right)} \quad (5.94)$$

$$= \left[2\mu^2 d(1 + \sinh 2\kappa d/2\kappa d)\right]^{-1}, \quad (5.95)$$

repeated use of (5.82) having being made in performing the last step of the calculation. We thus find, from (5.92), that

$$|R| = 1 \quad (5.96)$$

as expected.

Bibliographical notes

The expression for the reflection coefficient derived above was used by Linton and Evans (1991) to calculate approximate values for the frequencies at which trapped modes can exist above a submerged horizontal plate of finite length in water of constant depth. The essential idea, just as in §5.1.2, was to assume that the edges of the plate are sufficiently far apart so that there is little interaction between them. Martin (1995) showed how to take the awkward limit of $K_+(\mu)$ as $h \to \infty$ and thus recover results for infinite depth, found originally using the Wiener-Hopf technique by Greene and Heins (1953).

The problem in which the geometry is exactly as above but the wave is incident from the deep region was solved by Heins (1950b) who formulated the problem as an integral equation using Green's functions. McIver (1985a) reworked the problem using the simpler method due to Jones (1952).

5.2   Residue calculus theory

Many problems involving semi-infinite boundaries can be formulated in terms of eigenfunction expansions as described in Chapter 2. In such cases it is often possible to use the so-called residue calculus theory to obtain the explicit solution. The basic idea is to derive an infinite system of equations based on mode matching that can be solved exactly by constructing a meromorphic function f with an appropriate distribution of poles and zeros so that the unknowns in the infinite system correspond to the residues of f.

The method seems to have been first used by Whitehead (1951) and Berz (1951) in studies of the diffraction of microwaves. A number of papers presenting solutions using this technique to problems with semi-infinite boundaries appeared during the 1950s and 60s and the method is described in detail in Mittra and Lee (1971). For one particular problem, Mittra and Lee (§3-13) showed that the residue calculus technique is precisely equivalent to the Wiener-Hopf procedure. An attractive feature of the residue calculus theory is the ease with which it can be modified so as to produce numerically efficient methods for structures with finite boundaries. The method for dealing with finite length corrections seems to date back to VanBlaricum and Mittra (1969) and Itoh and Mittra (1969) and is described briefly in Jones (1994, §2.12). It is this aspect of the method that will be focussed on in the examples in §5.2.2 and §5.2.3.

In the application of the method it is necessary to use results concerning the asymptotics of infinite products and the relevant theory is given in §5.2.1 first.

5.2.1   Asymptotics of infinite products

In order to apply the residue calculus theory we will need the asymptotics of certain infinite products. In problems where $(\nabla^2 - \ell^2)\phi = 0$ is being solved subject to the standard bed and free-surface boundary conditions, the function

$$f(z) = \prod_{n=1}^{\infty} (1 - z/\alpha_n)\, e^{zh/n\pi} \qquad (5.97)$$

arises, where z is a complex variable, $\alpha_n = (k_n^2 + \ell^2)^{1/2}$ and k_n, $n = 1, 2, 3, \ldots$ are the positive real roots of the dispersion relation (2.5).

From (2.6) we have that $k_n h \sim n\pi + O(n^{-1})$ as $n \to \infty$, so that

$$\alpha_n = n\pi/h + O(n^{-1}). \tag{5.98}$$

This ensures that the terms in (5.97) are $1 + O(n^{-2})$ as $n \to \infty$ and hence that the product converges uniformly. To obtain asymptotics of this infinite product for large z we make use of the infinite product representation of the Gamma function (see Abramowitz and Stegun 1965, eqn 6.1.3),

$$\prod_{n=1}^{\infty} (1 - zh/n\pi) \, e^{zh/n\pi} = \frac{e^{\gamma zh/\pi}}{\Gamma(1 - zh/\pi)}, \tag{5.99}$$

in which $\gamma \approx 0.5772$ is Euler's constant. The asymptotics of the Gamma function are given by Stirling's formula (Abramowitz and Stegun 1965, eqn 6.1.37).

In order to determine the behaviour of $g(z)$ for large z, excluding the case when z is real and positive, we write

$$f(z) = \frac{e^{\gamma zh/\pi}}{\Gamma(1 - zh/\pi)} \prod_{n=1}^{\infty} \frac{\lambda_n(z - \alpha_n)}{\alpha_n(z - \lambda_n)}, \tag{5.100}$$

where $\lambda_n = n\pi/h$. Now

$$\prod_{n=1}^{\infty} \frac{\lambda_n(z - \alpha_n)}{\alpha_n(z - \lambda_n)} = \prod_{n=1}^{\infty} \frac{\lambda_n}{\alpha_n} \prod_{n=1}^{\infty} \left(1 + \frac{\lambda_n - \alpha_n}{z - \lambda_n}\right)$$

$$= \prod_{n=1}^{\infty} \frac{\lambda_n}{\alpha_n} \prod_{n=1}^{\infty} \left(1 + \frac{\lambda_n^2 - \alpha_n^2}{(z - \lambda_n)(\lambda_n + \alpha_n)}\right) \tag{5.101}$$

and because of the asymptotics of α_n for large n given by (5.98), both infinite products converge uniformly. Hence as $z \to \infty$, on a contour avoiding real, positive values,

$$\prod_{n=1}^{\infty} \frac{\lambda_n(z - \alpha_n)}{\alpha_n(z - \lambda_n)} \to \prod_{n=1}^{\infty} \frac{\lambda_n}{\alpha_n}, \tag{5.102}$$

so that

$$f(z) \sim e(z) \left(\frac{-1}{2zh}\right)^{1/2} \prod_{n=1}^{\infty} \frac{\lambda_n}{\alpha_n}, \tag{5.103}$$

where

$$e(z) = \exp\left[\frac{zh}{\pi}\left(\gamma - 1 + \ln\frac{-zh}{\pi}\right)\right].$$ (5.104)

Asymptotics for z real and positive can be obtained as follows. We note that

$$\prod_{n=1}^{\infty}(1 - z/\alpha_n)(1 + z/\alpha_n) = \frac{K\cos\zeta h + \zeta\sin\zeta h}{K(1 + \zeta^2/k^2)}\prod_{n=1}^{\infty}\frac{k_n^2}{\alpha_n^2},$$ (5.105)

where $\zeta = (z^2 - \ell^2)^{1/2}$, which follows from the Weierstrass product expansion (5.82). The product $\prod_{n=1}^{\infty} k_n^2/\alpha_n^2$ could also be evaluated explicitly with the aid of this identity, though this serves no purpose here. Hence, substituting (5.105) into (5.97) and then using (5.103) with z replaced by $-z$, we have that as $z \to \infty$ through real positive values

$$f(z) \sim (2zh)^{1/2}\frac{K\cos\zeta h + \zeta\sin\zeta h}{K(1 + \zeta^2/k^2)e(-z)}\prod_{n=1}^{\infty}\frac{k_n^2}{\lambda_n\alpha_n}.$$ (5.106)

For the case where the free-surface boundary condition is replaced by a Neumann condition, the product

$$g(z) = \prod_{n=1}^{\infty}(1 - z/\beta_n)\,e^{z/\lambda_n}$$ (5.107)

is encountered. Here $\beta_n = (\lambda_n^2 + \ell^2)^{1/2}$ and as $n \to \infty$, $\beta_n = \lambda_n + O(n^{-1})$. Asymptotics for this function can be derived exactly as above, or we can simply take the limit as $K \to 0$ in the above formulas, noting that in this limit $K \sim k^2 h$ and $k_n \to \lambda_n$.

It follows that as $z \to \infty$, provided z is not real and positive.

$$g(z) \sim \left(\frac{-\ell}{2z\sinh\ell h}\right)^{1/2}e(z),$$ (5.108)

and as $z \to \infty$ through real positive values,

$$g(z) \sim \frac{\sin\zeta h}{\zeta e(-z)}\left(\frac{2z\ell}{\sinh\ell h}\right)^{1/2}.$$ (5.109)

5.2.2 The finite dock problem

In §2.5.3 the problem of oblique wave scattering by a finite dock was formulated using matched eigenfunction expansions. One method of matching led to the pair of infinite systems of equations (2.138), which are reproduced here:

$$\sum_{n=0}^{\infty} V_n^{\pm} \left(\frac{1}{\alpha_n - \beta_m} \pm \frac{e^{-2\beta_m a}}{\alpha_n + \beta_m} \right) = \frac{1}{\alpha_0 + \beta_m} \pm \frac{e^{-2\beta_m a}}{\alpha_0 - \beta_m}, \qquad (5.110)$$

for $m = 0, 1, 2, \ldots$, where we require the unknowns to satisfy

$$V_n^{\pm} = O(n^{-1}) \quad \text{as} \quad n \to \infty \qquad (5.111)$$

so as to correctly model the singularity near the edges of the dock at $x = \pm a$, $z = 0$. The quantities which appear in (5.110) are defined by

$$\alpha_0 = -i\alpha = -i(k^2 - \ell^2)^{1/2}, \qquad \alpha_n = (k_n^2 + \ell^2)^{1/2}, \quad n \geq 1, \qquad (5.112)$$
$$\beta_n = (\lambda_n^2 + \ell^2)^{1/2}, \quad n \geq 0, \qquad (5.113)$$

where k and k_n are the solutions to (2.9) and (2.5), respectively, $\ell = k \sin \beta$ is the wavenumber in the y-direction (β being the angle with the positive x-axis made by the incident wave), and $\lambda_n = n\pi/h$. The reflection and transmission coefficients for the problem are given by

$$R = \frac{1}{2} \left(V_0^+ + V_0^- \right), \qquad T = \frac{1}{2} \left(V_0^+ - V_0^- \right) \qquad (5.114)$$

and here we will describe how these quantities can be determined using the residue calculus theory.

The system of equations (5.110) cannot be solved explicitly, but we can make use of the fact that if the exponential factors were not present, the system could be solved explicitly. This allows us to convert the problem into a system of equations which converges exponentially rather than algebraically. Thus we begin by solving the reduced system

$$\sum_{n=0}^{\infty} \frac{V_n}{\alpha_n - \beta_m} = \frac{1}{\alpha_0 + \beta_m}, \qquad m = 0, 1, 2, \ldots, \qquad (5.115)$$

for the unknown complex coefficients V_n. This is precisely the system of equations that is obtained from the matched eigenfunction expansion method for the semi-infinite dock problem, the dock occupying $x \geq 0$,

$z = 0$ and the reflection coefficient for that problem, which we will label as R_∞, is given by V_0.

Consider the function

$$g(z) = \frac{1}{z + \alpha_0} \prod_{n=0}^{\infty} \frac{1 - z/\beta_n}{1 - z/\alpha_n}. \tag{5.116}$$

It follows from (5.103), (5.106), (5.108), and (5.109) that

$$g(z) = O(z^{-1}) \quad \text{as} \quad z \to \infty \tag{5.117}$$

through a sequence of values which avoids the points $z = \alpha_n$ (which correspond to the points at which $K \cos \zeta h + \zeta \sin \zeta h = 0$).

If we consider the numbers

$$I_m = \lim_{N \to \infty} \frac{1}{2\pi i} \int_{C_N} \frac{g(z)}{z - \beta_m} \, dz, \qquad m = 0, 1, 2, \dots, \tag{5.118}$$

where C_N are circles centred on the origin with radius $(N + \frac{1}{2})\pi/h$, then the behaviour of $g(z)$ for large z implies that $I_m = 0$. Cauchy's residue theorem then shows that

$$\frac{R(g : -\alpha_0)}{-\alpha_0 - \beta_m} + \sum_{n=0}^{\infty} \frac{R(g : \alpha_n)}{\alpha_n - \beta_m} = 0, \tag{5.119}$$

for each $m \geq 0$, where $R(f : z_0)$ means the residue of $f(z)$ at $z = z_0$. If we write

$$G = \prod_{n=0}^{\infty} \frac{1 + \alpha_0/\alpha_n}{1 + \alpha_0/\beta_n}, \tag{5.120}$$

then

$$R(g : -\alpha_0) = G^{-1} \tag{5.121}$$

and so the solution of (5.115) is given by

$$V_n = G R(g : \alpha_n), \tag{5.122}$$

which can be shown, using techniques similar to those described in §5.2.1, to satisfy (5.111). In particular

$$V_0 = R_\infty = e^{-2i\beta} \prod_{n=1}^{\infty} \frac{(1 - i\alpha/\alpha_n)(1 + i\alpha/\beta_n)}{(1 + i\alpha/\alpha_n)(1 - i\alpha/\beta_n)} = e^{-2i\beta} e^{2i\delta_\infty}, \tag{5.123}$$

where

$$\delta_\infty = \sum_{n=1}^{\infty} \left(\tan^{-1}(\alpha/\beta_n) - \tan^{-1}(\alpha/\alpha_n) \right). \tag{5.124}$$

To treat the dock of finite extent we consider the function

$$f^\pm(z) = G^\pm g(z) h^\pm(z), \tag{5.125}$$

where $g(z)$ is given by (5.116) as before,

$$h^\pm(z) = 1 + \sum_{n=0}^{\infty} \frac{C_n^\pm}{z - \beta_n} \tag{5.126}$$

and $G^\pm$, $C_n^\pm$, $n \geq 0$, are constants to be determined. It is the inclusion of the function $g(z)$, which is used in the determination of the explicit solution to the semi-infinite dock problem, in the solution procedure for the finite dock problem that is crucial to the success of the technique. The numbers

$$I_m = \lim_{N \to \infty} \frac{1}{2\pi i} \int_{C_N} f^\pm(z) \left(\frac{1}{z - \beta_m} \pm \frac{e^{-2\beta_m a}}{z + \beta_m} \right) dz, \quad m = 0, 1, 2, \ldots, \tag{5.127}$$

are all zero for the same reasons as before and Cauchy's residue theorem gives

$$f^\pm(\beta_m) \pm e^{-2\beta_m a} f^\pm(-\beta_m) + \mathrm{R}(f^\pm : -\alpha_0) \left(\frac{1}{-\alpha_0 - \beta_m} \pm \frac{e^{-2\beta_m a}}{-\alpha_0 + \beta_m} \right)$$
$$+ \sum_{n=0}^{\infty} \mathrm{R}(f^\pm : \alpha_n) \left(\frac{1}{\alpha_n - \beta_m} \pm \frac{e^{-2\beta_m a}}{\alpha_n + \beta_m} \right) = 0, \tag{5.128}$$

for each $m \geq 0$. A comparison with (5.110) shows that the solution will be given by

$$V_n^\pm = \mathrm{R}(f^\pm : \alpha_n) \tag{5.129}$$

provided

$$\mathrm{R}(f^\pm : \alpha_n) = O(n^{-1}) \quad \text{as} \quad n \to \infty \tag{5.130}$$

(which follows from the fact that $f^\pm(z) = O(z^{-1})$ as $z \to \infty$), $G^\pm$ is chosen so that $\mathrm{R}(f^\pm : -\alpha_0) = 1$, and

$$f^\pm(\beta_m) \pm e^{-2\beta_m a} f^\pm(-\beta_m) = 0 \qquad m \geq 0. \tag{5.131}$$

We thus define

$$G^{\pm} = \frac{G}{h^{\pm}(-\alpha_0)}, \tag{5.132}$$

where G is defined in (5.120) and then provided (5.131) is satisfied we have

$$V_0^{\pm} = \mathrm{R}(f^{\pm} : \alpha_0) = \frac{h^{\pm}(-i\alpha)}{h^{\pm}(i\alpha)} R_{\infty}, \tag{5.133}$$

where R_{∞} is given by (5.123). In order to satisfy (5.131), the coefficients $C_m^{\pm}$, $m \geq 0$, used in (5.126) must satisfy the infinite systems of real equations

$$C_m^{\pm} \pm D_m \sum_{n=0}^{\infty} \frac{C_n^{\pm}}{\beta_m + \beta_n} = \pm D_m, \qquad m = 0, 1, 2, \ldots, \tag{5.134}$$

with

$$D_0 = 2\ell\, e^{-2\ell a} \prod_{n=1}^{\infty} \frac{(1 - \ell/\alpha_n)(1 + \ell/\beta_n)}{(1 + \ell/\alpha_n)(1 - \ell/\beta_n)}, \tag{5.135}$$

and for $m \geq 1$,

$$D_m = \frac{2\beta_m(\ell + \beta_m)(\alpha_m - \beta_m)}{(\ell - \beta_m)(\alpha_m + \beta_m)}\, e^{-2\beta_m a} \prod_{\substack{n=1 \\ n \neq m}}^{\infty} \frac{(1 - \beta_m/\alpha_n)(1 + \beta_m/\beta_n)}{(1 + \beta_m/\alpha_n)(1 - \beta_m/\beta_n)}. \tag{5.136}$$

If we combine the symmetric and antisymmetric solutions using (5.114) we find that the reflection and transmission coefficients are given by

$$R = \frac{1}{2}\left(e^{2i\delta_+} + e^{2i\delta_-}\right) R_{\infty}, \qquad T = \frac{1}{2}\left(e^{2i\delta_+} - e^{2i\delta_-}\right) R_{\infty}, \tag{5.137}$$

where

$$\delta^{\pm} = \arg\left(1 - \sum_{n=0}^{\infty} \frac{C_n^{\pm}}{i\alpha + \beta_n}\right). \tag{5.138}$$

As $a/h \to \infty$, (5.134) shows that $C_m^{\pm} \to 0$, and hence $\exp(2i\delta^{\pm}) \to 1$. Thus $R \to R_{\infty}$ and $T \to 0$ as expected.

Because of the presence of the factor $\exp(-2\beta_m a)$ in the expression (5.136) for D_m, the systems of equations (5.134) converge very quickly provided a/h is not too small and provide an extremely efficient method for computing the unknowns $C_m^\pm$. Moreover, it is possible to prove (using the method described in Appendix B of Evans 1992) that, for sufficiently large a/h, (5.134) has a unique solution with $\sum_{m=1}^\infty (C_m^\pm)^2 < \infty$.

Results for normal incidence can be extracted from the above analysis by taking the limit as $\beta \to 0$ (i.e. $\ell \to 0$ for fixed k). First we note that in this limit

$$D_0 \sim 2\ell + 4\ell^2(\sigma - a) + O(\ell^3), \tag{5.139}$$

where

$$\sigma = \sum_{n=1}^\infty \left(\frac{1}{\lambda_n} - \frac{1}{k_n} \right) \tag{5.140}$$

and hence the $m = 0$ equation in (5.134) with the plus sign then becomes $C_0^+ = 0$ and we only need to solve the system for $m \geq 1$. The leading order behaviour of the $m = 0$ equation in (5.134) with the minus sign is more complicated; we find that

$$(\sigma - a)C_0^- + \sum_{n=1}^\infty \frac{C_n^-}{\beta_n} = 1. \tag{5.141}$$

Apart from these changes, we can simply set $\ell = 0$ in the general expressions for R and T.

The procedure described above provides a numerically straightforward way of computing the reflection and transmission coefficients for the finite dock problem in finite depth. The infinite systems of equations that need to be solved converge extremely rapidly and the sums and products that need to be evaluated cause no difficulty. For example, the terms in the summation in the definition of δ_∞, equation (5.124), are $O(n^{-3})$ as $n \to \infty$. This is computationally acceptable, but the series is easily accelerated if necessary.

To demonstrate the rapid convergence of the infinite systems of equations we can consider the extreme case of a 1×1 truncation. If we only include one term from the summation in (5.134), solve for $C_0^\pm$, and substitute into (5.138) we obtain

$$\tan \delta^\pm \approx \frac{\pm \sin 2\beta}{b^{-1} \pm \cos 2\beta}, \tag{5.142}$$

		a/h		
β	0.25	0.5	0.75	1
$0°$	8.6	0.94	0.13	0.021
$40°$	5.2	0.43	0.046	0.0056
$80°$	0.61	0.026	0.0020	0.00020

Table 5.1 Maximum percentage error (over all frequencies) when computing $|R|$ for the finite dock problem using a 1×1 truncation.

where $b = D_0/2\ell$. If we substitute this expression into (5.137) and use (5.123) we obtain the approximations

$$R \approx e^{2i\delta_\infty} e^{2i\beta} \left(\frac{b^2 - 1}{b^2 - e^{4i\beta}} \right), \qquad T \approx b \, e^{2i\delta_\infty} \left(\frac{1 - e^{4i\beta}}{b^2 - e^{4i\beta}} \right). \qquad (5.143)$$

If we take the limit of (5.143) as $\beta \to 0$ using (5.139) we obtain the following approximations for the normal incidence case:

$$R \approx \frac{k(\sigma - a) \, e^{2i\delta_\infty}}{k(\sigma - a) - i}, \qquad T \approx \frac{-i \, e^{2i\delta_\infty}}{k(\sigma - a) - i}. \qquad (5.144)$$

The accuracy of these approximations depends strongly on the value of a/h and to a lesser extent on the value of β, with larger values of either parameter resulting in greater accuracy. This is illustrated in Table 5.1 which shows the errors that result from using these approximations to compute $|R|$. For each value of a/h and β the table gives the maximum percentage error in the computed value of $|R|$ as K varies over the entire frequency range. The table shows that 1% accuracy is achieved for all values of $a/h \geq 0.5$, with the accuracy increasing rapidly as a/h increases.

It is also possible to examine the long wave limit, i.e. $Kh \to 0$, for fixed β. In this limit $kh \sim (Kh)^{1/2}$ and an analysis of (5.134) reveals that

$$\delta^+ \sim \beta - ka \sin \beta \tan \beta, \qquad \delta^- \sim \beta - \frac{\pi}{2} + ka \cos \beta \qquad (5.145)$$

and hence that

$$R \sim -ika \sec \beta, \qquad T \sim 1 + ika \sec \beta \cos 2\beta. \qquad (5.146)$$

For $\beta = 0$, these results agree with those in Martin and Dalrymple (1988) after taking account of the different definition of T used in that paper.

Bibliographical notes

Problems concerning the interaction of water waves with a rigid plate lying in the free surface have a long history. The Wiener-Hopf technique was used in Heins (1948) to explicitly solve the interaction of oblique waves (generated by a line source along the edge of the plate) with a semi-infinite dock in water of finite depth. The case of normally incident waves is recovered by taking an appropriate limit. The method used is also applicable to infinite depth, but in that case the method breaks down in the limit corresponding to normally incident waves. However, this latter problem can be solved using complex variable theory and a method related to Laplace transforms (Friedrichs and Lewy 1948).

Most work on finite dock problems has concentrated on the infinite depth case and utilized an integral equation approach (some of which is described in the bibliographical notes at the end of §4.8). The difficult subject of short wave asymptotics for finite dock problems in infinite depth was the subject of a series of papers: Holford (1964a), (1964b), Leppington (1968), (1970), (1972), the last of these also including some results for finite water depths.

The first results for the finite depth case were based on shallow water theory (Stoker 1957) and then numerical calculations of the reflection and transmission coefficients were presented in Mei and Black (1969) for the scattering problem based on the full linear theory. A general numerical scheme, based on the finite-element method, for the solution of oblique scattering problems by infinite cylinders of constant cross-section lying in the free surface in water of finite depth was developed in Bai (1975). In particular, Bai considered the diffraction of waves by a cylinder with rectangular cross-section, including the case of zero draft, which corresponds to the finite dock problem. The finite depth problem has also been attacked using the complicated machinery of dual integral transforms in Dorfmann and Savvin (1998), though no numerical results were presented.

5.2.3 Periodic coastlines

Consider an ocean of constant depth h, bounded by a periodic coastline consisting of a straight, vertical cliff face from which protrudes an infinite number of equally-spaced identical thin vertical barriers, each one of length a and extending throughout the water depth. We will begin by assuming that a plane wave is incident on the coastline and calculate the scattered wave field. A plan view of the geometry is shown

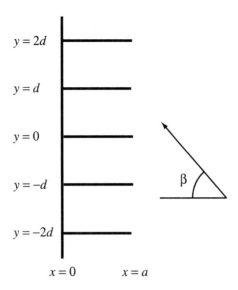

FIGURE 5.5
Scattering by a periodic coastline.

in Figure 5.5.

The depth variation can be factored out as in §2.4 and then the field equation to be solved is the two-dimensional Helmholtz equation

$$(\nabla^2 + k^2)\phi = 0 \tag{5.147}$$

in the fluid region, subject to the boundary conditions

$$\frac{\partial \phi}{\partial y} = 0 \quad \text{on} \quad x \in (0, a),\ y = md,\ m \text{ an integer}, \tag{5.148}$$

$$\frac{\partial \phi}{\partial x} = 0 \quad \text{on} \quad x = 0, \tag{5.149}$$

together with appropriate radiation conditions as $x \to \infty$.

The incident wave is described by

$$\phi_{\mathrm{I}} = e^{i\alpha a}\, e^{i\ell y - i\alpha x}, \tag{5.150}$$

where $\alpha = k\cos\beta$, $\ell = k\sin\beta$, and the phase factor $\exp(i\alpha a)$ has been introduced for convenience. As $\beta \in (0, \pi/2)$, it follows that $\ell \in (0, k)$. Since the incident wave is periodic in the y-direction and the array of

plates extends over the whole y-axis, we seek a scattered wave field which
has the term $\exp(i\ell y)$ in common with the incident wave. This, together
with the periodicity of the geometry, implies that the total potential
must satisfy

$$\phi(x, y + md) = e^{im\ell d}\,\phi(x, y) \tag{5.151}$$

and we only need consider the problem in the strip $y \in (0, d)$.

The problem is set up in a standard way by treating the two regions
$x \in (0, a)$ and $x \in (a, \infty)$ separately, expanding the potential in each of
these regions in terms of appropriate eigenfunctions, and then matching
ϕ and $\partial\phi/\partial x$ across $x = a$. Thus for $x \in (0, a)$ we write

$$\phi = \sum_{n=0}^{\infty} B_n \cos(n\pi y/d)\cosh\beta_n x, \tag{5.152}$$

where

$$\beta_n = (n^2\pi^2/d^2 - k^2)^{1/2} \tag{5.153}$$

and, for simplicity, we restrict attention to the case when only one wave
mode is possible between the plates, i.e. $kd < \pi$, in which case β_n, $n \geq 1$,
is real and $\beta_0 = -ik$. For $x \in (a, \infty)$ we write the wave field as

$$\phi = e^{i\ell y - i\alpha(x-a)} + \sum_{n=-\infty}^{\infty} A_n\, e^{i\ell_n y - \alpha_n(x-a)}, \tag{5.154}$$

where

$$\ell_n = \ell + 2n\pi/d \tag{5.155}$$

and

$$\alpha_0 = -i\alpha, \qquad \alpha_n = (\ell_n^2 - k^2)^{1/2}, \quad n \neq 0. \tag{5.156}$$

Since $\ell \in (0, k)$ and we have chosen $kd < \pi$, it follows that $|\ell_n| > k$,
$n \neq 0$. Hence α_n, $n \neq 0$, is real and there is just one reflected mode. As
$x \to \infty$,

$$\phi \sim e^{i\ell y - i\alpha(x-a)} + A_0\, e^{i\ell y + i\alpha(x-a)} \tag{5.157}$$

and thus the reflection coefficient is given by

$$R = A_0\, e^{-2i\alpha a}. \tag{5.158}$$

If we impose continuity of ϕ and $\partial\phi/\partial x$ across $x = a$, multiply each of the resulting equations by $\cos(m\pi y/d)$, $m = 0, 1, 2, \ldots$, integrate over $(0, d)$, and then eliminate B_n, we obtain an infinite system of equations for the unknowns A_n which can be written in the form

$$\sum_{n=-\infty}^{\infty} \frac{A_n \ell_n}{\ell} \left\{ \frac{1}{\alpha_n - \beta_m} + \frac{e^{-2\beta_m a}}{\alpha_n + \beta_m} \right\} = \frac{1}{\alpha_0 + \beta_m} + \frac{e^{-2\beta_m a}}{\alpha_0 - \beta_m}, \quad (5.159)$$

for $m = 0, 1, 2, \ldots$, which is of almost exactly the same form as (5.110). There are significant differences however. First, the infinite sum is over all integers n, not just non-negative n. Secondly, the definitions of α_n and β_n differ from those used in §5.2.2 in two crucial respects: β_0 is imaginary and, from (5.153), (5.155), and (5.156), we have $\alpha_n \sim n\pi/2d$, but $\beta_n \sim n\pi/d$ as $n \to \infty$. In order that the appropriate edge condition is satisfied (see equation 5.16) we require

$$A_n = O(n^{-3/2}) \quad \text{as} \quad n \to \infty. \quad (5.160)$$

In the finite dock problem considered in §5.2.2, we first solved a reduced system of equations in which terms which decayed exponentially as $a \to \infty$ were neglected. The effect of these neglected terms was then incorporated as a modification to the explicit solution of the reduced system. Here we will just examine the reduced system, though the finite length corrections (which would involve solving an exponentially convergent system of real equations) could easily be computed. For this scattering problem, unlike for the dock problem, the dependence on a is not totally absent in the reduced system. This is because β_0 is imaginary and so the term $\exp(-2\beta_0 a)$ does not decay as $a \to \infty$. Physically this corresponds to the fact that waves propagate in the regions between the barriers whatever the value of a, whereas in the dock problem there is only an evanescent wave-field beneath a semi-infinite dock.

Thus in order to obtain an approximate solution for large a/d we neglect the terms $\exp(-2\beta_m a)$, $m \geq 1$ and obtain

$$\sum_{n=-\infty}^{\infty} \frac{A_n \ell_n \ell^{-1}}{\alpha_n - \beta_m} = \frac{1}{\alpha_0 + \beta_m} + q\delta_{m0}, \quad m = 0, 1, 2, \ldots, \quad (5.161)$$

where

$$q = e^{2ika} \left[(\alpha_0 - \beta_0)^{-1} - C \right] \quad (5.162)$$

and

$$C = \sum_{n=-\infty}^{\infty} \frac{A_n \ell_n \ell^{-1}}{\alpha_n + \beta_0}. \tag{5.163}$$

This system of equations can be decomposed into two simpler systems by writing

$$A_n \ell_n \ell^{-1} = U_n + q V_n, \tag{5.164}$$

where, for $m = 0, 1, 2, \ldots$,

$$\sum_{n=-\infty}^{\infty} \frac{U_n}{\alpha_n - \beta_m} = \frac{1}{\alpha_0 + \beta_m}, \tag{5.165}$$

$$\sum_{n=-\infty}^{\infty} \frac{V_n}{\alpha_n - \beta_m} = \delta_{m0}. \tag{5.166}$$

The system (5.165) is similar to (5.115) which was solved in §5.2.2, but the required behaviour of the unknowns U_n is different since from (5.160) and (5.164) we must have $U_n, V_n = O(n^{-1/2})$ as $n \to \infty$.

Before solving these equations we shall examine the relationship between their solutions, U_n and V_n, and the reflection coefficient, which is related to A_0 through (5.158). From (5.163) and (5.164) we have

$$C = P + qQ, \tag{5.167}$$

where

$$P = \sum_{n=-\infty}^{\infty} \frac{U_n}{\alpha_n + \beta_0}, \qquad Q = \sum_{n=-\infty}^{\infty} \frac{V_n}{\alpha_n + \beta_0} \tag{5.168}$$

and elimination of C and q between (5.162), (5.164), and (5.167) shows that

$$A_0 = U_0 + V_0 \frac{(\alpha_0 - \beta_0)^{-1} - P}{e^{-2ika} + Q}. \tag{5.169}$$

In order to obtain the reflection coefficient, therefore, we need to know U_0 and V_0 together with the sums P and Q.

To solve (5.165) we consider the function

$$g(z) = G e^{-\frac{zd}{\pi} \ln 2} \frac{(1 - z/\beta_0)}{(1 - z^2/\alpha_0^2)} \prod_{n=1}^{\infty} \frac{(1 - z/\beta_n)}{(1 - z/\alpha_n)(1 - z/\alpha_{-n})}, \tag{5.170}$$

where G is a constant chosen so that $R(g : -\alpha_0) = 1$. The asymptotics of the product

$$\prod_{n=1}^{\infty} (1 - z/\beta_n) \, e^{zd/n\pi}$$

are given by (5.108) and (5.109) with h replaced by d and ℓ replaced by ik, and the asymptotics of the product

$$\prod_{n=1}^{\infty} (1 - z/\alpha_n)(1 - z/\alpha_{-n}) \, e^{zd/n\pi}$$

can be determined using the same techniques as were used to derive the results in §5.2.1, though the details are slightly different. From the resulting expressions we can show that

$$g(z) = O(z^{-1/2}) \quad \text{as} \quad z \to \infty \tag{5.171}$$

through a sequence of values which avoids the points $z = \alpha_n$, $n = 0, \pm 1, \pm 2, \dots$. Thus the numbers

$$I_m^{\pm} = \lim_{N \to \infty} \frac{1}{2\pi i} \int_{C_N} \frac{g(z)}{z \pm \beta_m} \, dz, \qquad m = 0, 1, 2, \dots, \tag{5.172}$$

where C_N are circles centred on the origin with radius $\ell_n + \pi/d$, are zero and Cauchy's residue theorem applied to I_m^- then gives

$$\sum_{n=-\infty}^{\infty} \frac{R(g : \alpha_n)}{\alpha_n - \beta_m} - \frac{1}{\alpha_0 + \beta_m} = 0, \qquad m = 0, 1, 2, \dots. \tag{5.173}$$

Comparison with (5.165) shows that

$$U_n = R(g : \alpha_n), \qquad n = 0, \pm 1, \pm 2 \dots, \tag{5.174}$$

which can be shown to satisfy $U_n = O(n^{-1/2})$ as $n \to \infty$ as required.

Cauchy's residue theorem applied to I_m^+ gives

$$\sum_{n=-\infty}^{\infty} \frac{R(g : \alpha_n)}{\alpha_n + \beta_m} - \frac{1}{\alpha_0 - \beta_m} + g(-\beta_m) = 0, \qquad m = 0, 1, 2, \dots, \tag{5.175}$$

and the $m = 0$ equation is simply, see (5.168),

$$P = (\alpha_0 - \beta_0)^{-1} - g(-\beta_0). \tag{5.176}$$

To solve (5.166) we consider the function

$$f(z) = \frac{F\,e^{-\frac{zd}{\pi}\ln 2}}{(1-z/\alpha_0)}\prod_{n=1}^{\infty}\frac{(1-z/\beta_n)}{(1-z/\alpha_n)(1-z/\alpha_{-n})},\qquad (5.177)$$

where F is a constant chosen so that $f(\beta_0) = -1$, and the numbers

$$J_m^{\pm} = \lim_{N\to\infty}\frac{1}{2\pi i}\int_{C_N}\frac{f(z)}{z\pm\beta_m}\,dz,\qquad m = 0,1,2,\dots, \qquad (5.178)$$

where C_N are as before. Cauchy's residue theorem implies

$$\sum_{n=-\infty}^{\infty}\frac{\mathrm{R}(f:\alpha_n)}{\alpha_n-\beta_m}-\delta_{m0}=0,\qquad m = 0,1,2,\dots, \qquad (5.179)$$

$$\sum_{n=-\infty}^{\infty}\frac{\mathrm{R}(f:\alpha_n)}{\alpha_n+\beta_m}+f(-\beta_m)=0,\qquad m = 0,1,2,\dots, \qquad (5.180)$$

and comparison with (5.166) and (5.168) shows that

$$V_n = \mathrm{R}(f:\alpha_n),\quad n = 0,\pm1,\pm2\dots, \qquad (5.181)$$
$$Q = -f(-\beta_0). \qquad (5.182)$$

Substitution of (5.174), (5.176), (5.181), and (5.182) into (5.169) gives

$$A_0 = \mathrm{R}(g:\alpha_0)\frac{e^{-2ika}-p^{-2}f(-\beta_0)}{e^{-2ika}-f(-\beta_0)}, \qquad (5.183)$$

where we have written

$$p = \frac{\beta_0-\alpha_0}{\beta_0+\alpha_0}=\frac{k-\alpha}{k+\alpha}=\frac{1-\cos\beta}{1+\cos\beta}=\tan^2\tfrac{1}{2}\beta \qquad (5.184)$$

and used the fact that

$$\mathrm{R}(f:\alpha_0)g(-\beta_0) = (1-p^{-2})\mathrm{R}(g:\alpha_0)f(-\beta_0). \qquad (5.185)$$

Simple calculations show that

$$f(-\beta_0) = p\,e^{2i\delta_1}, \qquad (5.186)$$
$$\mathrm{R}(g:\alpha_0) = -p\,e^{-2i\delta_2}, \qquad (5.187)$$

where

$$\delta_1 = -\frac{kd}{\pi}\ln 2 + \sum_{n=1}^{\infty}\left(\tan^{-1}\frac{k}{\alpha_n} + \tan^{-1}\frac{k}{\alpha_{-n}} - \tan^{-1}\frac{k}{\beta_n}\right), \quad (5.188)$$

$$\delta_2 = -\frac{\alpha d}{\pi}\ln 2 + \sum_{n=1}^{\infty}\left(\tan^{-1}\frac{\alpha}{\alpha_n} + \tan^{-1}\frac{\alpha}{\alpha_{-n}} - \tan^{-1}\frac{\alpha}{\beta_n}\right). \quad (5.189)$$

Combining (5.158), (5.183), (5.186), and (5.187) shows that the reflection coefficient

$$R = e^{-2i(\delta_2+\alpha a)}\,\frac{e^{2i(\delta_1+ka)} - p}{1 - p\,e^{2i(\delta_1+ka)}} \quad (5.190)$$

which satisfies $|R| = 1$ as expected, since all the incident wave energy is reflected by the coastline.

The accuracy of this result will depend on the value of a/d. Results from Linton and Evans (1993a), where this problem was used as part of the solution to a more complicated problem, suggest that for $a/d > 1$ the errors introduced by neglecting the exponentially decaying terms in (5.159) are very small indeed. As a/d decreases from unity, the accuracy gradually deteriorates with the best results when kd is small. This dependence of the error on kd is to be expected since the largest of the neglected exponentials is $\exp(-2(a/d)(\pi^2 - k^2 d^2)^{1/2})$ which will become more significant as kd approaches π.

In the above analysis we have assumed that $\ell < k$ but, just as for the case of a submerged horizontal cylinder considered in §3.1.1, it is possible to treat the case $\ell > k$ and look for frequencies at which there exist non-trivial solutions of the homogeneous problem obtained in the absence of the incident wave. For $\ell > k$ waves are no longer possible as $x \to \infty$ and any such non-trivial solutions that can be found correspond to edge waves that travel along the coastline but do not radiate any energy out to sea, or equivalently to Rayleigh-Bloch surface waves propagating along a diffraction grating.

The system of equations that we must solve when $\ell > k$ is the same as (5.159) except with zero on the right-hand side and we note that in this case α_n is real for all integers n. Once again we will neglect the exponentially decaying terms and if we proceed much as before we find that we must again solve (5.166), the solution being given by (5.181), but now our solution must satisfy the condition

$$f(ik) = e^{-2ika}. \quad (5.191)$$

From (5.186), but noting that now

$$p = \frac{ik + \alpha_0}{ik - \alpha_0} = -\exp(2i\tan^{-1} k/\alpha_0), \qquad (5.192)$$

this condition reduces to

$$\tan^{-1}(k/\alpha_0) + \delta_1 + ka = (m + \tfrac{1}{2})\pi, \qquad (5.193)$$

for some integer m. This approximate condition for the existence of edge waves was first derived, using this technique, by Hurd (1954). The extension to include the exponentially decaying terms, which also demonstrates the high accuracy of (5.193), was carried out in Evans and Linton (1993a).

Bibliographical notes

The scattering problem considered here is an example of wave scattering by a diffraction grating, a subject with a vast literature in the context of acoustic and electromagnetic waves; Petit (1980) gives a general discussion of such problems, while Wilcox (1984) provides a more mathematical approach to the general theory. See also the notes at the end of §6.4.3.

When the barriers extend all the way to $x = -\infty$, the problem can be solved explicitly. For various electromagnetic diffraction problems this was done using the Wiener-Hopf technique by Carlson and Heins (1947), Heins and Carlson (1947), Heins (1950a) and using residue calculus theory by Whitehead (1951), though in all these papers the more general problem of scattering by an array of staggered semi-infinite plates was considered. The transmission of sound through a staggered array of finite length plates was treated using a modification of the Wiener-Hopf technique in Koch (1971). For the case of zero stagger described above, the solution of the semi-infinite plate problem essentially boils down to solving (5.165), the solution of which is given by (5.174).

The question of the existence of modes propagating along a general periodic grating in the absence of any wave field away from the grating is still an open one, though they are known not to exist if the boundary condition on the grating is $\phi = 0$ rather than $\partial\phi/\partial n = 0$ (Wilcox 1984, Bonnet-Bendhia and Starling 1994). Such Rayleigh-Bloch surface waves have, however, been observed experimentally by Barlow and Karbowiak (1954) for the case of a corrugated cylindrical metal surface, and they also showed how the frequencies of such modes could be associated with the zeros of an infinite determinant. Rayleigh-Bloch waves have been constructed for a number of other geometries. For example Evans and Fernyhough (1995) calculated the frequencies at which such modes occur when the thin barriers in our example are replaced by barriers of non-zero thickness. Computations for general shapes have been made, using an integral equation formulation, by Porter and Evans (1999).

Chapter 6

Arrays

Wave interactions between neighbouring structures are important in many practical problems, including the scattering of water waves by an offshore oil platform and the scattering of acoustic waves by a tube bundle in a heat exchanger. In this chapter a variety of methods are presented for such interaction problems.

The problem of wave diffraction by a finite array of circular cylinders in which the fluid motion is governed by the two-dimensional Helmholtz equation has a long history, and a straightforward and efficient method of solution for this problem is presented in §6.1. In the water-wave context this cylinder problem corresponds to a structure extending throughout the depth. For the case when the cylinder is truncated, so that there is fluid beneath and/or above it, the situation is considerably more complex as a doubly-infinite set of additional modes are required to describe the wave field. An efficient method for this more general problem is described in §6.2.

In view of the complexity of interaction problems involving many structures it is natural to seek approximations to the solution. A very effective procedure is to assume that the structures are widely spaced in comparison to appropriate length scales. Such 'wide-spacing' approximations have been found to perform well even when the assumptions behind them are violated. In §6.3 a method of this type for the solution of two-dimensional water-wave problems is described. For the special case of scattering by an array of equally-spaced identical structures exact expressions for the overall reflection and transmission coefficients may be found and this is demonstrated in §6.3.1, while in §6.3.2 the extension is made to an infinite two-dimensional array.

A theory for the diffraction of waves by an infinite row, or grating, of circular cylinders has already been presented in §3.2.2. The wide-spacing

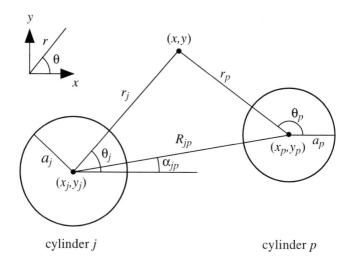

FIGURE 6.1
Plan view of two cylinders.

approximation is used in §6.4 to obtain a method applicable to multiple diffraction gratings. Reciprocity relations for multiple diffraction gratings are derived in §6.4.2.

6.1   An array of vertical circular cylinders

We begin by tackling the problem of wave scattering by a finite array of fixed vertical circular cylinders extending throughout the water depth. We assume that there are N (≥ 1) cylinders and use $N+1$ polar coordinate systems in the (x, y)-plane: (r, θ) centred at the origin and (r_j, θ_j), $j = 1, \ldots, N$, centred at (x_j, y_j), the centre of the jth cylinder. The various parameters relating to the relative positions and sizes of the cylinders are shown in Figure 6.1. An extension of the method to the case of an infinite row of cylinders (the same example as was considered using the multipole method described in §3.2.2) will be considered later.

As in §3.2.2 we assume that a plane wave making an angle β with the positive x-axis is incident on the cylinders and the depth dependence of the problem is factored out so that the problem is reduced to solving the Helmholtz equation (2.74) in the (x, y)-plane. The incident wave is characterized by

$$\phi_I = \exp(i\alpha x + i\ell y) = \exp(ikr\cos(\theta - \beta)), \tag{6.1}$$

where $\alpha = k\cos\beta$ and $\ell = k\sin\beta$.

A phase factor for each cylinder, I_j, is defined by

$$I_j = e^{i\alpha x_j + i\ell y_j} \tag{6.2}$$

and then we can write

$$\phi_I = I_j\, e^{ikr_j\cos(\theta_j - \beta)} = I_j \sum_{n=-\infty}^{\infty} e^{in(\pi/2 - \theta_j + \beta)}\, J_n(kr_j), \tag{6.3}$$

(Gradshteyn and Ryzhik 1980, eqn 8.511(4)).

Each cylinder will scatter the waves which are incident upon it and we can take account of all such scatterings by expressing the total potential as the sum of the incident wave and a general outgoing wave emanating from each cylinder. Thus (see equation 2.67), the total potential is written

$$\phi = \phi_I + \sum_{j=1}^{N} \sum_{n=-\infty}^{\infty} A_n^j Z_n^j H_n^{(1)}(kr_j)\, e^{in\theta_j}, \tag{6.4}$$

for some set of unknown complex coefficients A_n^j and the factor $Z_n^j = J_n'(ka_j)/H_n^{(1)'}(ka_j)$ is introduced for convenience.

In order to apply the boundary condition of no flow through the cylinder surfaces, which is

$$\frac{\partial \phi}{\partial r_p} = 0 \quad \text{on} \quad r_p = a_p, \quad p = 1, \dots, N, \tag{6.5}$$

equation (6.4) must be written solely in terms of the coordinates r_p and θ_p. This is accomplished using Graf's addition theorem for Bessel functions (Gradshteyn and Ryzhik 1980, eqn 8.530), which shows that provided $r_p < R_{jp}$ for all j, we can write

$$\phi(r_p, \theta_p) = \sum_{n=-\infty}^{\infty} \left(I_p J_n(kr_p)\, e^{in(\pi/2 - \theta_p + \beta)} + A_n^p Z_n^p H_n^{(1)}(kr_p)\, e^{in\theta_p} \right)$$

$$+ \sum_{\substack{j=1 \\ \neq p}}^{N} \sum_{n=-\infty}^{\infty} A_n^j Z_n^j \sum_{m=-\infty}^{\infty} J_m(kr_p) H_{n-m}^{(1)}(kR_{jp})\, e^{im\theta_p}\, e^{i(n-m)\alpha_{jp}}. \tag{6.6}$$

The geometrical restriction implies that this expression is only valid if the point (r_p, θ_p) is closer to the centre of cylinder p than the centres of any of the other cylinders. This is certainly true on the surface of cylinder p and so (6.6) can be used to apply the body boundary condition which leads, after using the orthogonality of the functions $\exp(im\theta_p)$, $m = 0, \pm1, \pm2, \ldots$, to the system of equations

$$A^p_m + \sum_{\substack{j=1 \\ \neq p}}^{N} \sum_{n=-\infty}^{\infty} A^j_n Z^j_n \, e^{i(n-m)\alpha_{jp}} \, H^{(1)}_{n-m}(kR_{jp}) = -I_p \, e^{im(\pi/2-\beta)},$$

$$p = 1, \ldots, N, \quad m = 0, 1, 2, \ldots . \quad (6.7)$$

Equation (6.7) can be substituted back into (6.6) and it follows that provided $r_p < R_{jp}$ for all j,

$$\phi(r_p, \theta_p) = \sum_{n=-\infty}^{\infty} A^p_n \, e^{in\theta_p} \left(Z^p_n H^{(1)}_n(kr_p) - J_n(kr_p) \right). \quad (6.8)$$

This expression provides an extremely simple formula for the velocity potential, and hence the free-surface elevation, in the vicinity of a cylinder. The potential on the cylinder surfaces has a particularly simple form since, using (A.7), we have

$$\phi(a_p, \theta_p) = -\frac{2i}{\pi k a_p} \sum_{n=-\infty}^{\infty} \frac{A^p_n \, e^{in\theta_p}}{H^{(1)'}_n(ka_p)}. \quad (6.9)$$

If we put $N = 1$ in (6.7) and take the cylinder to be at the origin with $\beta = 0$, we find that $A^1_m = -i^m$ and we recover the results for a single cylinder given in §2.4.1.

As an example of the sort of results that this theory predicts, Figure 6.2 shows the results of computations of the exciting force (defined in equation 1.45) on four identical cylinders arranged in a square. This example is the same as that used in Figure 2 of Linton and Evans (1990), but the curves shown in that paper are in error. The values have been non-dimensionalized by the forces that would be experienced if the cylinders were in isolation, so the curves represent the effects of the interaction due to the array and they clearly demonstrate that interaction effects can be extremely important in determining the amplitude of the force. For example, the force on cylinder 1 near $ka = 1.7$ is over twice what it would be if the other cylinders were not present.

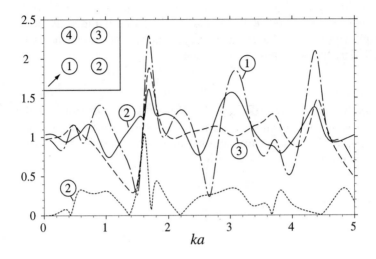

FIGURE 6.2
The non-dimensional amplitude of the forces on a group of four cylinders arranged in a square. The cylinders, numbered 1–4 are at $(-h, -h)$, $(h, -h)$, (h, h), $(-h, h)$, respectively, and in each case the radius is $h/2$. The incident wave makes an angle of $45°$ with the positive x-axis. The upper curves show the force in the direction of wave advance on cylinders 1, 2, and 3 (the magnitude of the force on cylinder 4 is identical to that on cylinder 2) whereas the lower curve shows the force in the direction perpendicular to wave-advance on cylinder 2 (this is zero on cylinders 1 and 3).

The same approach as has been used above for the finite array case can be used to treat the case of an infinite row of identical bottom-mounted vertical circular cylinders which was solved using the multipole method in §3.2.2. Thus we can consider a row of identical cylinders of radius a arranged so that the centres of the cylinder cross-sections are at $(x, y) = (0, 2md)$, $m = 0, \pm1, \pm2, \ldots$. The total potential must now be written as

$$\phi = \phi_I + \sum_{m=-\infty}^{\infty} \sum_{n=-\infty}^{\infty} A_n^m Z_n H_n^{(1)}(kr_m)\, e^{in\theta_m}, \qquad (6.10)$$

The quantities A_n^m are unknown complex numbers which must be found by applying the boundary condition on the cylinders. Due to the

periodicity of the geometry it is clear that the only difference between
the effect of the mth cylinder and that of the cylinder situated at the
origin will be that due to the different phase of the incident wave at that
cylinder. Thus we have

$$A_n^m = I_m A_n \tag{6.11}$$

where we have written A_n for A_n^0. With this considerable simplification
we now need only apply the body boundary condition on one of the
cylinders. This leads (in much the same way as for the finite array case)
to the infinite system of equations

$$A_p + \sum_{n=-\infty}^{\infty} A_n Z_n \, \mathrm{e}^{\mathrm{i}(p-n)\pi/2} \sum_{m=1}^{\infty} H_{n-p}^{(1)}(2mkd) \left[I_m + (-1)^{n-p} I_{-m} \right]$$
$$= - \mathrm{e}^{\mathrm{i}p(\pi/2-\beta)}, \qquad p = 0, \pm 1, \pm 2, \ldots . \tag{6.12}$$

The sum over m is slowly convergent and must be considered carefully
in any numerical computations. Methods for the accurate evaluation of
sums of the form $\sum_{m=1}^{\infty} c_m H_n^{(1)}(mx)$ have been considered by a number of previous authors, see for example Yeung and Sphaier (1989) and
Thomas (1991).

The system of equations (6.12) can be used to derive a compact expression for ϕ near to the cylinders. We obtain, for $a < r < 2d$,

$$\phi(r,\theta) = \sum_{n=-\infty}^{\infty} A_n \, \mathrm{e}^{\mathrm{i}n\theta} \left(Z_n H_n^{(1)}(kr) - J_n(kr) \right) \tag{6.13}$$

and

$$\phi(a,\theta) = -\frac{2\mathrm{i}}{\pi ka} \sum_{n=-\infty}^{\infty} \frac{A_n \, \mathrm{e}^{\mathrm{i}n\theta}}{H_n^{(1)'}(ka)} \tag{6.14}$$

which are of precisely the same form as (6.8) and (6.9) derived above for
the finite array case, though the unknown coefficients are the solutions
of a different system of equations.

One disadvantage of solving this particular problem in the manner
described above is that the nature of the far field is not immediately
apparent. The potential in (6.10) is expressed as an infinite sum of
outgoing circular waves, but as $|x| \to \infty$ we know that the solution
must behave like a sum of plane waves. The reflection and transmission
coefficients, which determine the amplitudes of these far-field waves,

are thus not readily evaluated from this formulation, but comparison of (6.13) with (3.95) shows that the unknown coefficients A_n are related to the unknowns a_n and b_n used in the multipole solution procedure, by

$$a_n = \frac{\epsilon_n i^n}{2} (A_n + A_{-n}) Z_n, \tag{6.15}$$

$$b_n = i^{n+1} (-A_n + A_{-n}) Z_n. \tag{6.16}$$

Thus a_n and b_n can be evaluated and R_p and T_p computed from (3.99)–(3.101).

Bibliographical notes

The method of solution for finite arrays described in this section was devised by Záviška (1913) in the context of two-dimensional acoustics and the system of equations (6.7) was derived by Row (1955) (using a technique based on integral equations) for the problem in which the boundary condition on the cylinders is $\phi = 0$. The first application of the method to the water wave problem was carried out by Spring and Monkmeyer (1974) and the simplification leading to (6.8) was first noted by Linton and Evans (1990). The same method was used by Kim (1993) to solve the much more complicated radiation problem for an array of cylinders in an arbitrary mode of motion.

An approximate solution to this problem was devised by McIver and Evans (1984a) based on the work of Simon (1982). If the cylinders are assumed to be widely spaced it is possible to approximate the circular waves that emanate from one cylinder as plane waves when they encounter the other cylinders. This 'plane-wave approximation' actually produces accurate results even for closely-spaced cylinders, but does not lead to significant savings in effort.

Záviška's method was applied to the case of normal incidence on an infinite row of cylinders by von Ignatowsky (1914) and was extended to oblique incidence in Twersky (1962). The normal incidence problem was solved in the context of water waves by Spring and Monkmeyer (1975).

6.2 A general interaction theory

It is possible to derive an interaction theory for a much more general class of structures than the bottom-mounted vertical circular cylinders considered in §6.1. Kagemoto and Yue (1986) showed how this could be done for an array of structures in water of constant finite depth, which have the property that they are 'vertically separated'. The precise meaning of this restriction will be made clear following (6.20) below, but arrays containing structures with intersecting vertical projections are certainly excluded. One structural geometry for which the method of Kagemoto and Yue (1986) is suitable is the truncated cylinder that has already been discussed in §2.5.2.

Unlike for the case of bottom-mounted wall-sided structures such as the cylinder array in §6.1, the depth dependence cannot be factored out from this less restrictive problem and so the general form for an outgoing cylindrical wave emanating from the jth structure is written (see equation 2.62)

$$\phi_{\text{S}}^j(r_j, \theta_j, z) = \sum_{m=0}^{\infty} \psi_m(z) \sum_{n=-\infty}^{\infty} A_{mn}^j K_n(k_m r_j) \, e^{in\theta_j} \, . \tag{6.17}$$

Here (r_j, θ_j, z) are cylindrical polar coordinates, with origin O_j fixed in the undisturbed free surface, and axis $r_j = 0$ passing through the jth structure. The distance between the local origins O_j and O_p is R_{jp} and the relative position of the origins is determined by the angles α_{jp} as shown in Figure 6.1. The depth eigenfunctions $\psi_m(z)$ and the numbers k_m are defined in §2.1 and we note that $k_0 = -ik$, where k is the positive solution to the dispersion relation $k \tanh kh = K$. The $m = 0$ terms in the expansion thus correspond (see equation A.5) to outgoing cylindrical waves, whereas (see equation A.9) the terms with $m \geq 1$ represent evanescent modes, which decay exponentially away from the structure.

For the vertical circular cylinders considered previously, an infinite system of algebraic equations was derived which could be solved numerically by truncation. Here we proceed slightly differently by truncating the infinite summations at the outset. Thus we introduce truncation parameters M and N_m, $m = 0, 1, \ldots, M$ and write

$$\phi_{\text{S}}^j(r_j, \theta_j, z) = \sum_{m=0}^{M} \psi_m(z) \sum_{n=-N_m}^{N_m} A_{mn}^j K_n(k_m r_j) \, e^{in\theta_j} \tag{6.18}$$

$$= \boldsymbol{A}_j^T \boldsymbol{\Psi}_j(r_j, \theta_j, z), \tag{6.19}$$

where $\boldsymbol{A}_j$ is the vector of coefficients A_{mn}^j and $\boldsymbol{\Psi}_j$ is the vector of scattered partial cylindrical waves $\psi_m(z) K_n(k_m r_j) \exp(in\theta_j)$. In Kagemoto and Yue's original treatment, the $m = 0$ terms were considered separately, but this is unnecessary provided we always bear in mind that $K_n(k_0 r)$ and $I_n(k_0 r)$ are related to $H_n^{(1)}(kr)$ and $J_n(kr)$ through (A.5) and (A.4), respectively. The precise ordering of the elements A_{mn}^j in the vector $\boldsymbol{A}_j$ and the partial waves $\psi_m(z) K_n(k_m r_j) \exp(in\theta_j)$ in the vector $\boldsymbol{\Psi}_j$ is unimportant, provided of course that they are consistent.

It follows from Graf's addition theorem (Gradshteyn and Ryzhik 1980, eqn 8.530) that, for $j, p = 1, \ldots, N$, $j \neq p$,

$$K_n(k_m r_j) \, e^{in(\theta_j - \alpha_{jp})} = \sum_{l=-\infty}^{\infty} K_{n+l}(k_m R_{jp}) I_l(k_m r_p) \, e^{il(\pi - \theta_p + \alpha_{jp})},$$

(6.20)

provided $r_p < R_{jp}$ for all j. (The use of this result means that, in addition to the requirement for non-overlapping vertical sections, which is necessary for (6.17) to be valid in the immediate neighbourhood of a structure, we must also have an array for which the escribed vertical circular cylinder to each structure, centred at its respective origin, does not enclose any other origin.) Thus, using a suitably truncated version of (6.20), we can write

$$\boldsymbol{\Psi}_j = \boldsymbol{T}_{jp} \boldsymbol{\Phi}_p,$$

(6.21)

where $\boldsymbol{\Phi}_p$ is a vector of 'incident' partial cylindrical waves of the form $\psi_m(z) I_n(k_m r_p) \exp(in\theta_p)$ and the element of the matrix $\boldsymbol{T}_{jp}$ which relates the scattered partial wave $\psi_m(z) K_n(k_m r_j) \exp(in\theta_j)$ to the incident partial wave $\psi_m(z) I_l(k_m r_p) \exp(il\theta_p)$ is given by

$$[\boldsymbol{T}_{jp}]_{nl} = (-1)^l K_{n-l}(k_m R_{jp}) \, e^{i(n-l)\alpha_{jp}} \, .$$

(6.22)

A combination of (6.19) and (6.21) shows that the potential ϕ_S^j can be evaluated in terms of the wave field at structure p as

$$\phi_S^j(r_p, \theta_p, z) = \boldsymbol{A}_j^T \boldsymbol{T}_{jp} \boldsymbol{\Phi}_p$$

(6.23)

and so the total field incident upon structure p is

$$\chi_I^p = \phi_I^p + \sum_{\substack{j=1 \\ \neq p}}^{N} \boldsymbol{A}_j^T \boldsymbol{T}_{jp} \boldsymbol{\Phi}_p,$$

(6.24)

where ϕ_I^p is the incident wave written in terms of (r_p, θ_p, z). From (6.3) this is just

$$\phi_I^p = I_p \psi_0(z) \sum_{n=-N_0}^{N_0} (-1)^n I_n(k_0 r_p) \, e^{in(\theta_p - \beta)} = \boldsymbol{a}_p^T \boldsymbol{\Phi}_p, \qquad (6.25)$$

where $\boldsymbol{a}_p$ is the vector of coefficients of the partial wave decomposition of the incident wave about O_p, all of whose components are zero except those corresponding to the propagating modes. Thus

$$\chi_I^p = \left(\boldsymbol{a}_p^T + \sum_{\substack{j=1 \\ \neq p}}^N \boldsymbol{A}_j^T \boldsymbol{T}_{jp} \right) \boldsymbol{\Phi}_p. \qquad (6.26)$$

At this point in the development of the theory it is instructive to look back at the special case of bottom-mounted cylinders considered in §6.1. In that case (6.6) gives the total field in the vicinity of cylinder p and is made up of three parts. The first term in the initial sum is just ϕ_I^p and the final triple summation is equivalent to the sum in (6.26). The remaining part is a general outgoing wave from cylinder p and the application of the boundary condition on that cylinder leads to the system of equations (6.7). This system relates the coefficients in the expansion of the wave emanating from cylinder p to the coefficients in the expansion of the incident field at p (as in equation 6.26) through the boundary condition on cylinder p.

In general, it is possible to relate the incident and diffracted partial waves at the pth structure through the diffraction characteristics of that structure in isolation. Thus there exist 'diffraction transfer matrices' $\boldsymbol{B}_p$, $p = 1, \ldots, N$ (often referred to as T-matrices, see Martin 1985b) such that, from (6.26),

$$\boldsymbol{A}_p = \boldsymbol{B}_p \left(\boldsymbol{a}_p^T + \sum_{\substack{j=1 \\ \neq p}}^N \boldsymbol{A}_j^T \boldsymbol{T}_{jp} \right)^T. \qquad (6.27)$$

Specifically, the element $[\boldsymbol{B}_p]_{nl}$ is the amplitude of the nth partial wave of the scattered potential due to a wave of mode l incident on body p in isolation.

In order to use this interaction theory therefore, we must first obtain the elements of $\boldsymbol{B}_p$, $p = 1, \ldots, N$, and then the unknown amplitudes $\boldsymbol{A}_p$ can be determined from the system of equations (6.27). For general geometries the matrices $\boldsymbol{B}_p$ are fully populated, but for axisymmetric structures they are sparse, which simplifies the procedure somewhat. Thus for an array of axisymmetric structures we can consider, for each p, the scattering of an incident field of the form $\psi_m(z)I_n(k_m r_p)\exp(in\theta_p)$, $m = 0, 1, \ldots, M$, $n = 0, \pm 1, \pm 2, \ldots, \pm N_m$, by the pth structure in isolation, with a resulting diffracted field of the form

$$\sum_{l=0}^{M} c_{lmn}^{p} \psi_l(z) K_n(k_l r_p) \, e^{il\theta_p} \,.$$

The coefficients c_{lmn}^{p} are the elements of $\boldsymbol{B}_p$.

For the bottom-mounted circular cylinders considered in §6.1 we only need to consider the case $m = 0$ and the factor $\psi_0(z)$ can be ignored. An incident field of the form $I_n(k_0 r_p)\exp(in\theta_p)$ results in a scattered field $c_n^p K_n(k_0 r_p)\exp(in\theta_p)$ and, from the results in §2.4.1 and bearing (A.4) and (A.5) in mind, we have that for this case $\boldsymbol{B}_p$ is diagonal and

$$[\boldsymbol{B}_p]_{nn} = c_n^p = \frac{2\mathrm{i}}{\pi}(-1)^n Z_n^p, \tag{6.28}$$

where $Z_n^p = J_n'(ka_p)/H_n^{(1)\prime}(ka_p)$. From (6.25) it follows that

$$[\boldsymbol{a}_p]_n = I_p(-1)^n \, e^{-in\beta} \tag{6.29}$$

and a comparison of (6.4) and (6.17) shows that

$$[\boldsymbol{A}_p]_n = \frac{2}{\pi}(-\mathrm{i})^{n+1} A_n^p Z_n^p. \tag{6.30}$$

The elements of $\boldsymbol{T}_{jp}$ are given in (6.22) and if the values from (6.28)–(6.30) are substituted into (6.27) we obtain the infinite system (6.7) as before.

In general it is not possible to determine the elements of the matrices $\boldsymbol{B}_p$ analytically and instead they must be determined numerically, perhaps using one of the many techniques described in this book. We have only described the scattering problem here, but the extension to radiation problems is straightforward.

Bibliographical notes

Multiple scattering problems have a long history and a vast associated literature; some examples have already been discussed in §3.2. In the context of electromagnetic waves a technique for solving such problems was described by Heaviside (1893). His approach was an iterative one in which each structure diffracts waves towards the others, leading to further scattered fields, each of which is then scattered by all the structures, and so on. In the context of acoustics this procedure was pioneered by Twersky (1952) and it has been used in water-wave problems by, for example, Ohkusu (1974), Greenhow (1980), Duncan and Brown (1982), Mavrakos and Koumoutsakos (1987), and Mavrakos (1991). This technique shares with the method described in this section the advantage that the multiple scattering problem is solved in terms of the (assumed known) solutions to the individual scattering problems, but it rapidly becomes unmanageable as the number of scatterers increases.

The method of Kagemoto and Yue (1986) was used by Yilmaz and Incecik (1998) and Yilmaz (1998) to solve the problems of scattering and radiation by an array of truncated vertical circular cylinders, and it has been modified to cope with very large arrays by Kashiwagi (1999) and Chakrabarti (2000). Both these latter authors use the idea of dividing the array into identical modules, each of which contains many structures, and computing the characteristics of an individual module before combining the modules into the full array.

An approximate theory for the scattering and radiation of waves by an array of structures that are each small relative to the wavelength was developed in the context of wave-power devices by Kyllingstad (1984).

6.3   The wide-spacing approximation

Here a 'wide-spacing approximation' is developed for two-dimensional water-wave problems involving a number of structures. The basis of the approximation is to assume that the structures are sufficiently widely spaced for evanescent modes to be neglected when calculating the interactions between neighbouring structures. Martin (1985a) examined the approximation in detail and concluded that it is based on the assumptions that the spacings between structures are large compared to both the wavelength and the size of each structure. Thus, if L is a typical spacing, a a typical size, and k the wavenumber, then the approximation formally requires $kL \gg 1$ and $L/a \gg 1$.

Consider N, not necessarily identical, structures in the vicinity of $x = L_n$, $n = 1, 2, \ldots, N$. The hydrodynamic theory described in §1.3

was developed for a single structure; the extension to N structures is straightforward and may be considered as the addition of extra modes of motion. In two dimensions there are three possible modes of motion (heave, sway, and roll) for a single structure and hence there are a total of $3N$ modes of motion for the array. The hydrodynamics of the array can be described fully by consideration of the scattering problem, when the structures are all held fixed, and a sequence of radiation problems in which each structure in turn is forced to oscillate in a single mode of motion while the remaining structures are held fixed. Both the scattering and radiation problems will be included within a single formulation.

To construct the wide-spacing approximation the scattering and radiation properties of the isolated individual structures are needed. When structure n is forced to oscillate in mode μ then the corresponding potential has the form

$$\phi_{n\mu} \sim A_{n\mu}^{\pm} \, e^{\pm ik(x-L_n)} \, \frac{\cosh k(z+h)}{\cosh kh} \quad \text{as} \quad x - L_n \to \pm\infty \qquad (6.31)$$

(cf. equation 1.51). For a wave incident from the left, the reflection and transmission coefficients for structure n are denoted by R_{n1} and T_{n1}, respectively, and for a wave incident from the right by R_{n2} and T_{n2}. From (1.61) it is known that $T_{n1} = T_{n2}$, but the distinction is retained in the following so that the origin of terms can more easily be identified.

To facilitate the solution, the fluid region is divided into $N+1$ regions as follows:

$$\begin{array}{lll} \text{Region } 1: & x \in (-\infty, L_1), & \\ \text{Region } n: & x \in (L_{n-1}, L_n), & n = 2, 3, \ldots, N, \\ \text{Region } N+1: & x \in (L_N, \infty). & \end{array} \qquad (6.32)$$

In region n, far enough from the structures for evanescent waves to be negligible, the potential is written as a combination of plane waves so that

$$\phi \approx \left[A_n \, e^{ik(x-L_n)} + B_n \, e^{-ik(x-L_n)} \right] \frac{\cosh k(z+h)}{\cosh kh},$$

$$n = 1, 2, \ldots, N+1. \qquad (6.33)$$

It is assumed that a wave of amplitude A is incident from the left of the array (to obtain a radiation problem A is set to zero) and no wave is incident from the right; thus

$$A_1 = -\frac{igA}{\omega} \, e^{ikL_1} \quad \text{and} \quad B_{N+1} = 0 \qquad (6.34)$$

(cf. equation 1.17). In addition it is assumed that at most only structure m oscillates in a single mode of motion to produce outgoing waves as in (6.31); by the linearity of the problem combinations of motions are obtained by superposition.

In region $n+1$ the representation of the potential is

$$
\begin{aligned}
\phi &\approx \left[A_{n+1}\, e^{ik(x-L_{n+1})} + B_{n+1}\, e^{-ik(x-L_{n+1})} \right] \frac{\cosh k(z+h)}{\cosh kh} \\
&= \left[A_{n+1}\, e^{-ik\Delta_n}\, e^{ik(x-L_n)} + B_{n+1}\, e^{ik\Delta_n}\, e^{-ik(x-L_n)} \right] \frac{\cosh k(z+h)}{\cosh kh},
\end{aligned}
\tag{6.35}
$$

where $\Delta_n = L_{n+1} - L_n$. Note that L_{N+1} does not correspond to the position of any structure, but it is mathematically convenient to include it here. The wave propagating away from structure n in region $n+1$ arises from the transmission of the wave propagating toward the structure in region n, the reflection of the wave propagating toward the structure in region $n + 1$, and the wave radiated away from the structure in region $n + 1$; thus

$$
A_{n+1}\, e^{-ik\Delta_n} = T_{n1} A_n + R_{n2} B_{n+1} + \delta_{mn} A_{n\mu}^{+},
\tag{6.36}
$$

where δ_{mn} is the Kronecker delta. Similarly, the wave propagating away from structure n in region n arises from the transmission of the wave propagating toward the structure in region $n + 1$, the reflection of the wave propagating toward the structure in region n, and the wave radiated away from the structure in region n; thus

$$
B_n = T_{n2} B_{n+1}\, e^{ik\Delta_n} + R_{n1} A_n + \delta_{mn} A_{n\mu}^{-}.
\tag{6.37}
$$

In matrix notation, the above equations are

$$
\begin{aligned}
\begin{pmatrix} e^{-ik\Delta_n} & -R_{n2}\, e^{ik\Delta_n} \\ 0 & T_{n2}\, e^{ik\Delta_n} \end{pmatrix} \begin{pmatrix} A_{n+1} \\ B_{n+1} \end{pmatrix} \\
= \begin{pmatrix} T_{n1} & 0 \\ -R_{n1} & 1 \end{pmatrix} \begin{pmatrix} A_n \\ B_n \end{pmatrix} + \delta_{mn} \begin{pmatrix} A_{n\mu}^{+} \\ -A_{n\mu}^{-} \end{pmatrix}
\end{aligned}
\tag{6.38}
$$

or, after inversion of the first matrix,

$$
\begin{pmatrix} A_{n+1} \\ B_{n+1} \end{pmatrix} = S_n \begin{pmatrix} A_n \\ B_n \end{pmatrix} + \delta_{mn} S'_n \begin{pmatrix} A_{n\mu}^{+} \\ -A_{n\mu}^{-} \end{pmatrix},
\tag{6.39}
$$

where the so-called scattering matrix

$$S_n = \frac{1}{T_{n2}} \begin{pmatrix} [T_{n1}T_{n2} - R_{n1}R_{n2}] \, e^{ik\Delta_n} & R_{n2} \, e^{ik\Delta_n} \\ -R_{n1} \, e^{-ik\Delta_n} & e^{-ik\Delta_n} \end{pmatrix} \tag{6.40}$$

and

$$S'_n = \frac{1}{T_{n2}} \begin{pmatrix} T_{n2} \, e^{ik\Delta_n} & R_{n2} \, e^{ik\Delta_n} \\ 0 & e^{-ik\Delta_n} \end{pmatrix}. \tag{6.41}$$

Note that

$$\det S_n = \frac{T_{n1}}{T_{n2}} = 1 \tag{6.42}$$

by virtue of (1.61).

Successive applications of (6.39) lead to

$$\begin{pmatrix} A_{N+1} \\ 0 \end{pmatrix} = P_N \begin{pmatrix} -\frac{igA}{\omega} \, e^{ikL_1} \\ B_1 \end{pmatrix} + P_{N-m} S'_m \begin{pmatrix} A^+_{m\mu} \\ -A^-_{m\mu} \end{pmatrix}, \tag{6.43}$$

where

$$P_0 = I_2, \quad P_n = S_N S_{N-1} \dots S_{N-n+1} \ (n = 1, 2, \dots, N), \tag{6.44}$$

and I_2 denotes the 2×2 identity matrix. Given the scattering and radiation properties of an individual structure, equation (6.43) can be solved to determine B_1 and A_{N+1} which are proportional to the amplitudes of the waves radiated to infinity on either side of the array. The wave amplitudes in any of the regions may then be found from (6.39) and hence the hydrodynamic forces on any of the structures calculated. For structure n in isolation, let $\hat{X}^\pm_{n\mu}$ be the exciting force in direction μ when the structure is held fixed in a wave of unit amplitude incident from $x = \pm\infty$, and let $f_{n\mu}$ be the force due to the forced oscillations of the structure in mode μ in the absence of any incident wave (cf. equations 1.45 and 1.46). The total hydrodynamic force on the structure is

$$F_{n\mu} = \frac{i\omega}{g} A_n \hat{X}^-_{n\mu} + \frac{i\omega}{g} B_{n+1} \hat{X}^+_{n\mu} + \delta_{mn} f_{n\mu}. \tag{6.45}$$

6.3.1   Scattering by equally-spaced identical structures

To illustrate further the application of the wide-spacing approximation the scattering by N identical, equally-spaced structures will be examined in more detail. The subscript n is dropped in the reflection and transmission coefficients and the spacing between structures is denoted by L. The scattering matrix (6.40) then becomes

$$
\mathbf{S} = \frac{1}{T_2} \begin{pmatrix} [T_1 T_2 - R_1 R_2]\, \mathrm{e}^{\mathrm{i}kL} & R_2\, \mathrm{e}^{\mathrm{i}kL} \\ -R_1\, \mathrm{e}^{-\mathrm{i}kL} & \mathrm{e}^{-\mathrm{i}kL} \end{pmatrix} \tag{6.46}
$$

or, in terms of the modulus and phase introduced in equation (1.64),

$$
\mathbf{S} = \frac{1}{|T|} \begin{pmatrix} \mathrm{e}^{\mathrm{i}(\delta+kL)} & |R|\, \mathrm{e}^{\mathrm{i}(\delta_2-\delta+kL)} \\ -|R|\, \mathrm{e}^{\mathrm{i}(\delta_1-\delta-kL)} & \mathrm{e}^{-\mathrm{i}(\delta+kL)} \end{pmatrix} \equiv \begin{pmatrix} s_{11} & s_{12} \\ s_{21} & s_{22} \end{pmatrix}, \tag{6.47}
$$

where (1.62) has been used to simplify one of the components of $\mathbf{S}$. In this scattering problem, in addition to (6.34),

$$
B_1 = -\frac{\mathrm{i}gA}{\omega} \widetilde{R}_N\, \mathrm{e}^{-\mathrm{i}kL_1} \quad \text{and} \quad A_{N+1} = -\frac{\mathrm{i}gA}{\omega} \widetilde{T}_N\, \mathrm{e}^{\mathrm{i}kL_{N+1}}, \tag{6.48}
$$

where $\widetilde{R}_N$ and $\widetilde{T}_N$ are, respectively, the reflection and transmission coefficients for the complete array. With these values (6.43) reduces to

$$
\begin{pmatrix} \widetilde{T}_N\, \mathrm{e}^{\mathrm{i}k(L_N+L)} \\ 0 \end{pmatrix} = \mathbf{S}^N \begin{pmatrix} \mathrm{e}^{\mathrm{i}kL_1} \\ \widetilde{R}_N\, \mathrm{e}^{-\mathrm{i}kL_1} \end{pmatrix} \tag{6.49}
$$

and if

$$
\mathbf{S}^N = \begin{pmatrix} t_{11} & t_{12} \\ t_{21} & t_{22} \end{pmatrix} \tag{6.50}
$$

is known then it follows that

$$
\widetilde{R}_N = -\frac{t_{21}}{t_{22}}\, \mathrm{e}^{2\mathrm{i}kL_1} \quad \text{and} \quad \widetilde{T}_N = \mathrm{e}^{\mathrm{i}k(L_1-L_N-L)}\left(t_{11} - \frac{t_{12}t_{21}}{t_{22}}\right). \tag{6.51}
$$

The matrix power $\mathbf{S}^N$ may be evaluated numerically, but it is also possible to obtain an explicit expression by using a result from elementary matrix algebra that gives the power in terms of the eigenvalues and eigenvectors of the matrix (Edwards and Penney 1988, §6.3). The eigenvalues of $\mathbf{S}$ are $\mathrm{e}^{\pm\alpha}$ where

$$
\cosh\alpha = \frac{\cos(\delta+kL)}{|T|} \equiv f(kL) \tag{6.52}
$$

and $f(kL)$ is real for all kL. If $|f(kL)| \leq 1$ then

$$\alpha = i \cos^{-1} f(kL) \tag{6.53}$$

is pure imaginary, while if $|f(kL)| > 1$ then

$$\alpha = \cosh^{-1} |f(kL)| + ip\pi, \tag{6.54}$$

where p is an even (odd) integer when $f(kL)$ is positive (negative). After some manipulation it is found that the elements of S^N are given by

$$
\begin{aligned}
t_{11} &= (s_{11} \sinh N\alpha - \sinh(N-1)\alpha)/\sinh\alpha, \\
t_{12} &= s_{12} \sinh N\alpha/\sinh\alpha, \\
t_{21} &= -\sinh N\alpha (e^{\alpha} - s_{11})(e^{-\alpha} - s_{11})/(s_{12} \sinh\alpha), \\
t_{22} &= (\sinh(N+1)\alpha - s_{11} \sinh N\alpha)/\sinh\alpha,
\end{aligned}
\tag{6.55}
$$

where there are the restrictions that $|R| \neq 0$ and $|T| \neq 0$ for the particular kL. Evans (1990) gives an equivalent form for S^N which he verifies by induction. From (6.51) it then follows that

$$\widetilde{T}_N = \frac{e^{ik(L_1 - L_N - L)} \sinh\alpha}{\sinh(N+1)\alpha - s_{11} \sinh N\alpha}. \tag{6.56}$$

When $|f(kL)| > 1$, so that α has a non-zero real part, then $\widetilde{T}_N \to 0$ as $N \to \infty$ so that there are ranges of frequency for which transmission is blocked by the array. This is discussed further in §6.3.2.

Figure 6.3 compares the wide-spacing approximation with an accurate solution for two surface-piercing vertical barriers. Both the results for a single vertical barrier, required for the wide-spacing approximation, and the results for two barriers were calculated by the method of Porter and Evans (1995). The barriers are submerged to a depth a in water of depth h and are a distance L apart. The figure shows the modulus of the reflection coefficient $|R|$ as a function of the wavenumber kL for $a/h = 0.5$ and $L/a = 2$. An assumption of the wide-spacing approximation is that $kL \gg 1$, but the accuracy of the approximation is good even for modest kL. A second assumption is that $L/a \gg 1$ and it can be seen that for $L/a = 2$ there is little deviation from the accurate solution; for $L/a = 4$ (not shown) the two methods are graphically indistinguishable.

6.3.2 Wave propagation through a periodic array

The wide-spacing approximation may be used to investigate the propagation of waves through a periodic array of scatterers that extends to

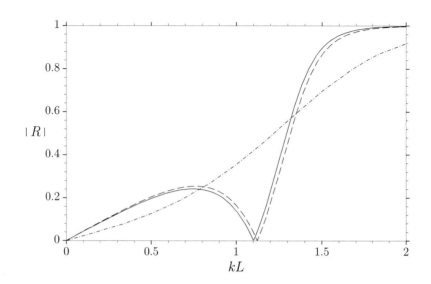

FIGURE 6.3
The reflection coefficient for two vertical barriers calculated by
the wide-spacing approximation ($- - - -$). Comparison is made
with an accurate solution (————) for two vertical barriers
and the solution for a single vertical barrier with the same
submergence ($- \cdot - \cdot - \cdot -$).

infinity in both horizontal directions. Suppose that there are identical
scatterers at $x = nL$ for all integers n. Solutions are sought in the form

$$\phi(x, z) = e^{iqx} \psi(x, z), \qquad (6.57)$$

where the complex wave number q is to be found, and ψ has the same
periodicity as the array so that

$$\psi(x + L, z) = \psi(x, z). \qquad (6.58)$$

Such solutions are often called Bloch waves and the problem is referred
to here as the Bloch problem. The above is equivalent to looking for
solutions in the form

$$\phi(x + L, z) = e^{iqL} \phi(x, z) \qquad (6.59)$$

so that moving one period through the array scales the potential by
a factor $T_{\mathrm{B}} = e^{iqL}$ which is sometimes called the Bloch transmission

coefficient. The real part of q measures the change in phase as the wave propagates. If q has a non-zero imaginary part then there is also a change in amplitude.

To determine the possible forms of q it is sufficient to consider a single period of length L and thus attention is restricted to $x \in [-L/2, L/2]$. Equation (6.59) is equivalent to the two independent periodicity conditions

$$\phi(L/2, z) = e^{iqL} \phi(-L/2, z)$$
$$\frac{\partial \phi}{\partial x}(L/2, z) = e^{iqL} \frac{\partial \phi}{\partial x}(-L/2, z). \tag{6.60}$$

Under the assumptions of the wide-spacing approximation, in the vicinity of $x = -L/2$ the potential

$$\phi \approx \left(A_1 e^{ikx} + B_1 e^{-ikx}\right) \frac{\cosh k(z+h)}{\cosh kh} \tag{6.61}$$

while in the vicinity of $x = -L/2$

$$\phi \approx \left(A_2 e^{ikx} + B_2 e^{-ikx}\right) \frac{\cosh k(z+h)}{\cosh kh}. \tag{6.62}$$

Similar arguments to those used to arrive at equations (6.36) and (6.37) lead to

$$A_2 = T_1 A_1 + R_2 B_2 \quad \text{and} \quad B_1 = T_2 B_2 + R_1 A_1, \tag{6.63}$$

while the Bloch conditions (6.60) yield

$$A_2 = A_1 e^{i(q-k)L} \quad \text{and} \quad B_2 = B_1 e^{i(q+k)L}. \tag{6.64}$$

Elimination of A_2 and B_2 gives

$$S \begin{pmatrix} A_1 \\ B_1 \end{pmatrix} = T_B \begin{pmatrix} A_1 \\ B_1 \end{pmatrix} \tag{6.65}$$

so that the Bloch transmission coefficient T_B is given by the eigenvalues of the scattering matrix S given in (6.46). Thus

$$T_B = e^{\pm\alpha}, \tag{6.66}$$

where α is given by (6.52). When α is pure imaginary this corresponds to waves that propagate through the array without change of amplitude and this occurs for ranges of frequency known as passing bands. When

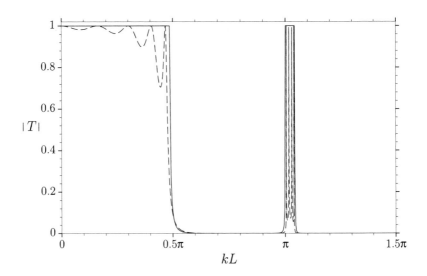

FIGURE 6.4
The transmission coefficient $|T|$ vs. kL for a finite array of vertical barriers as calculated by the wide-spacing approximation
($- - - -$) and the Bloch solution (———).

α has a non-zero real part, the two possible solutions have an amplitude that either grows or decays as the wave propagates; this occurs for ranges of frequency known as stopping bands.

The Bloch transmission coefficient may be used to obtain an estimate of the transmission through a finite array. In Figure 6.4 a comparison for the case of $N = 5$ barriers is made between the transmission coefficient $|T_N|$, as calculated from the wide-spacing approximation (6.56), and $|T_B|^N$ where the Bloch solution corresponding to decaying waves is selected. The geometrical parameters are the same as for Figure 6.3. The passing bands for the Bloch solution correspond to $|T_B| = 1$. For most of the stopping bands the transmission coefficient is very close to zero. The transmission coefficient $|T_5|$ for five barriers broadly follows the Bloch solution and, in particular, the Bloch solution predicts very well when transmission is blocked by the finite array. As noted after (6.56), as $N \to \infty$ in the finite array then $T_N \to 0$ for exactly the values of kL which correspond to stopping bands in the Bloch problem.

Bibliographical notes

The extension of many of the the reciprocity relations discussed in §1.4 to an array of structures is given by Srokosz and Evans (1979) for two dimensions and by Evans (1980) and Falnes (1980) for three dimensions.

The wide-spacing approximation for the scattering of waves by two fixed obstacles is described by Newman (1965). The extension to include both scattering and radiation problems for two structures was made by Srokosz and Evans (1979) in the context of wave-power devices. A rather thorough investigation of the scattering problem for an arbitrary number of identical, equally-spaced obstacles is given by Evans (1990) (the latter work also discusses the application of the wide-spacing approximation to sloshing problems). Similar formulations arise in other contexts, for instance the propagation of water waves over ripples in the sea bed (see, for example, Porter and Chamberlain 1997).

A wide-spacing approximation method for three-dimensional water-wave problems was developed by Simon (1982) and refined by McIver and Evans (1984a), who solved the problem described in §6.1 using the technique. This so-called 'plane-wave approximation' was also used by Abul-Azm and Williams (1989) to solve the scattering problem for an array of truncated circular cylinders, but for three-dimensional problems the approximate technique seems to have little or no computational advantage over the general formulations described in §§6.1–6.2.

The Bloch wave problem is well known in solid-state physics and Ashcroft and Mermin (1976) give a detailed account of the formalism that has been developed. In the context of water waves, Chou (1998) applied the theory to wave interaction with surface scatterers. The wide-spacing approximation was applied to the Bloch problem by Heckl (1992) in the context of sound propagation through tube bundles.

6.4 Diffraction by multiple gratings

Here the wide-spacing approximation is developed further to include the scattering of waves by a finite number of diffraction gratings. An individual grating consists of an infinite row of identical vertical cylinders lying along a line of constant x and with a reference axis within each cylinder a horizontal distance $2d$ from its nearest neighbour in the row. An example of such a grating was considered in §3.2.2 where the geome-

try is an infinite row of vertical circular cylinders. The depth dependence
is removed in the manner described in §2.4 so that all potentials satisfy
the Helmholtz equation (2.74) in the x, y-plane.

First of all, in §6.4.1 a description of the diffraction of waves by a single
grating is given and this is followed in §6.4.2 by a derivation of some
relations between involving the reflection and transmission matrices for
a diffraction grating. The application of the wide-spacing approximation
to multiple diffraction gratings is given in §6.4.3.

6.4.1   A single grating

Before considering multiple gratings, the properties of a single grating
are detailed using a modification of the notation used in §3.2.2. Suppose
that this single grating lies along $x = 0$. A wave with potential

$$\phi_I = e^{i\alpha x + i\ell y} = e^{ikr\cos(\theta - \theta_q)} \tag{6.67}$$

is incident from the left at an angle θ_q to the x-axis, where k is the
wavenumber,

$$\alpha = k\cos\theta_q, \quad \ell = k\sin\theta_q, \tag{6.68}$$

and (r, θ) are polar coordinates defined by $(x, y) = (r\cos\theta, r\sin\theta)$. This
wave will be diffracted to obtain a reflected field

$$\phi_D \sim \sum_{p=-\mu}^{\nu} R_{qp}^{(1)} e^{-ikr\cos(\theta + \theta_p)} \quad \text{as} \quad x \to -\infty \tag{6.69}$$

and a transmitted field

$$\phi_D \sim \sum_{p=-\mu}^{\nu} T_{qp}^{(1)} e^{ikr\cos(\theta - \theta_p)} \quad \text{as} \quad x \to \infty, \tag{6.70}$$

where

$$\sin\theta_p = \sin\theta_q + p\pi/kd, \tag{6.71}$$

$$\mu = [(1 + \sin\theta_q)kd/\pi], \quad \nu = [(1 - \sin\theta_q)kd/\pi], \tag{6.72}$$

and $[\cdot]$ indicates that the integer part should be taken. The set of angles
defined by (6.71) is closed in the sense that the far-field forms (6.69)–
(6.70) are valid for any incident wave direction θ_q, with integer $q \in [-\mu, \nu]$.

For a wave incident from the right with potential

$$\phi_{\mathrm{I}} = e^{-i\alpha x + i\ell y} = e^{-ikr\cos(\theta+\theta_q)} \tag{6.73}$$

the reflected field is

$$\phi_{\mathrm{D}} \sim \sum_{p=-\mu}^{\nu} R_{qp}^{(2)} \, e^{ikr\cos(\theta-\theta_p)} \quad \text{as} \quad x \to \infty \tag{6.74}$$

and the transmitted field is

$$\phi_{\mathrm{D}} \sim \sum_{p=-\mu}^{\nu} T_{qp}^{(2)} \, e^{-ikr\cos(\theta+\theta_p)} \quad \text{as} \quad x \to -\infty. \tag{6.75}$$

If there is symmetry of the grating about $x = 0$ then $R_{qp}^{(2)} = R_{qp}^{(1)}$ and $T_{qp}^{(2)} = T_{qp}^{(1)}$ for all p and q.

In the generalization of the theory to multiple gratings, an incident wave made up of all components allowed by (6.71)–(6.72) must be allowed for. To this end, it is convenient to introduce some further notation. Let $\boldsymbol{u}$ and $\boldsymbol{v}$ be the vectors with components

$$u_q = e^{ikr\cos(\theta-\theta_q)} \quad \text{and} \quad v_q = e^{-ikr\cos(\theta+\theta_q)}, \quad q = -\mu, \dots, \nu, \tag{6.76}$$

respectively. Thus, the components of $\boldsymbol{u}$ are rightward propagating waves and the components of $\boldsymbol{v}$ are leftward propagating waves. Suppose a wave with potential

$$\phi_{\mathrm{I}} = \sum_{q=-\mu}^{\nu} a_q u_q = \boldsymbol{a}^T \boldsymbol{u} \tag{6.77}$$

is incident upon the grating from the left, where the components of $\boldsymbol{a}$ are assumed to be known, then from (6.69)–(6.70) the total reflected field

$$\phi_{\mathrm{D}} \sim \sum_{p,q=-\mu}^{\nu} a_q R_{qp}^{(1)} v_p = \boldsymbol{a}^T \boldsymbol{R}^{(1)} \boldsymbol{v} \quad \text{as} \quad x \to -\infty \tag{6.78}$$

and the total transmitted field

$$\phi_{\mathrm{D}} \sim \sum_{p,q=-\mu}^{\nu} a_q T_{qp}^{(1)} u_p = \boldsymbol{a}^T \boldsymbol{T}^{(1)} \boldsymbol{u} \quad \text{as} \quad x \to \infty. \tag{6.79}$$

Here $\boldsymbol{R}^{(i)}$ and $\boldsymbol{T}^{(i)}$ are the matrices with components $R_{qp}^{(i)}$ and $T_{qp}^{(i)}$ respectively. Similarly, for an incident wave

$$\phi_I = \sum_{q=-\mu}^{\nu} a_q v_q = \boldsymbol{a}^T \boldsymbol{v} \tag{6.80}$$

from the right, (6.74)–(6.75) give the reflected field

$$\phi \sim \sum_{p,q=-\mu}^{\nu} a_q R_{qp}^{(2)} u_p = \boldsymbol{a}^T \boldsymbol{R}^{(2)} \boldsymbol{u} \quad \text{as} \quad x \to \infty \tag{6.81}$$

and the transmitted field

$$\phi \sim \sum_{p,q=-\mu}^{\nu} a_q T_{qp}^{(2)} v_p = \boldsymbol{a}^T \boldsymbol{T}^{(2)} \boldsymbol{v} \quad \text{as} \quad x \to -\infty. \tag{6.82}$$

6.4.2   Reciprocity relations

Here various relations are derived that involve the amplitudes of the reflected and transmitted waves in a diffraction problem for a grating. The derivations use applications of Green's theorem in a similar way to those employed in §1.4.

Consider the two scattering potentials $\phi_i(x, y)$, $i = 1, 2$ satisfying

$$\phi_i(x, y) \sim \begin{cases} \boldsymbol{c}_i^T \boldsymbol{u} + \boldsymbol{d}_i^T \boldsymbol{v}, & x \to -\infty, \\ \boldsymbol{e}_i^T \boldsymbol{u} + \boldsymbol{f}_i^T \boldsymbol{v}, & x \to \infty. \end{cases} \tag{6.83}$$

By virtue of the y dependence in the wave components, such potentials satisfy the conditions

$$\phi_i(x, d) = e^{2i\ell d} \phi_i(x, -d) \tag{6.84}$$

and

$$\frac{\partial \phi_i}{\partial y}(x, d) = e^{2i\ell d} \frac{\partial \phi_i}{\partial y}(x, -d); \tag{6.85}$$

see equations (3.75) and (3.76).

Reciprocity relations are obtained by applications of Green's theorem around a rectangular contour consisting of the lines $x = \pm X$, $|y| < d$, and the lines $y = \pm d$, $|x| < X$, where $X > 0$. In general, for the contribution from $y = d$ to cancel that from $y = -d$, ϕ_1 and $\overline{\phi}_2$ (or $\overline{\phi}_1$ and ϕ_2) must be used in Green's theorem and hence

$$\int_\Gamma \left(\phi_1 \frac{\partial \overline{\phi}_2}{\partial n} - \overline{\phi}_2 \frac{\partial \phi_1}{\partial n} \right) \, \mathrm{d}s = 0, \tag{6.86}$$

where Γ consists of the lines at $x = \pm X$. In the limit $X \to \infty$, application of the far-field forms (6.83) then gives

$$\boldsymbol{c}_1^T \boldsymbol{W} \overline{\boldsymbol{c}}_2 - \boldsymbol{d}_1^T \boldsymbol{W} \overline{\boldsymbol{d}}_2 = \boldsymbol{e}_1^T \boldsymbol{W} \overline{\boldsymbol{e}}_2 - \boldsymbol{f}_1^T \boldsymbol{W} \overline{\boldsymbol{f}}_2, \tag{6.87}$$

where $\boldsymbol{W}$ is the diagonal matrix with non-zero entries $\cos\theta_q$, $q = -\mu, \dots, \nu$, and the over bar denotes complex conjugate.

Define left and right scattering potentials ϕ_L and ϕ_R, respectively, where

$$\phi_\mathrm{L} \sim \begin{cases} \boldsymbol{a}^T(\boldsymbol{u} + \boldsymbol{R}^{(1)}\boldsymbol{v}), & x \to -\infty, \\ \boldsymbol{a}^T \boldsymbol{T}^{(1)} \boldsymbol{u}, & x \to \infty \end{cases} \tag{6.88}$$

and

$$\phi_\mathrm{R} \sim \begin{cases} \boldsymbol{a}^T \boldsymbol{T}^{(2)} \boldsymbol{v}, & x \to -\infty, \\ \boldsymbol{a}^T(\boldsymbol{v} + \boldsymbol{R}^{(2)}\boldsymbol{u}), & x \to \infty. \end{cases} \tag{6.89}$$

The forms (6.88)–(6.89) effectively define the wave amplitudes in (6.83) that are appropriate to ϕ_L and ϕ_R. Application of (6.87) gives the following:

(i) $\phi_1 = \phi_2 = \phi_\mathrm{L} \Rightarrow$

$$\boldsymbol{R}^{(1)} \boldsymbol{W} \boldsymbol{R}^{(1)^*} + \boldsymbol{T}^{(1)} \boldsymbol{W} \boldsymbol{T}^{(1)^*} = \boldsymbol{W}, \tag{6.90}$$

(ii) $\phi_1 = \phi_2 = \phi_\mathrm{R} \Rightarrow$

$$\boldsymbol{R}^{(2)} \boldsymbol{W} \boldsymbol{R}^{(2)^*} + \boldsymbol{T}^{(2)} \boldsymbol{W} \boldsymbol{T}^{(2)^*} = \boldsymbol{W}, \tag{6.91}$$

(iii) $\phi_1 = \phi_\mathrm{L}, \phi_2 = \phi_\mathrm{R} \Rightarrow$

$$\boldsymbol{R}^{(1)} \boldsymbol{W} \boldsymbol{T}^{(2)^*} + \boldsymbol{T}^{(1)} \boldsymbol{W} \boldsymbol{R}^{(2)^*} = \boldsymbol{0}, \tag{6.92}$$

(iv) $\phi_1 = \phi_\mathrm{R}, \phi_2 = \phi_\mathrm{L} \Rightarrow$

$$\boldsymbol{R}^{(2)} \boldsymbol{W} \boldsymbol{T}^{(1)^*} + \boldsymbol{T}^{(2)} \boldsymbol{W} \boldsymbol{R}^{(1)^*} = \boldsymbol{0}. \tag{6.93}$$

Here $*$ denotes conjugate transpose. A reduction of equations (6.90)–(6.93) for the case of a symmetric grating was obtained by Twersky (1962). Equations (6.90)–(6.91) express conservation of energy flux and for a symmetric grating are equivalent to (3.103). The above may be written in block matrix form as

$$\begin{pmatrix} R^{(1)} & T^{(1)} \\ T^{(2)} & R^{(2)} \end{pmatrix} \begin{pmatrix} W & 0 \\ 0 & W \end{pmatrix} \begin{pmatrix} R^{(1)*} & T^{(2)*} \\ T^{(1)*} & R^{(2)*} \end{pmatrix} = \begin{pmatrix} W & 0 \\ 0 & W \end{pmatrix}. \tag{6.94}$$

For some special cases of the boundary conditions (6.84)–(6.85) it is possible to apply Green's theorem to a pair of functions ϕ_i, $i = 1, 2$, for which the contributions to the boundary integral from $y = \pm d$ vanish individually. For example, this happens if either homogeneous Neumann or Dirichlet conditions are applied on $y = \pm d$ to obtain a wave guide problem. In these cases, in addition to (6.86), the relation

$$\int_\Gamma \left(\phi_1 \frac{\partial \phi_2}{\partial n} - \phi_2 \frac{\partial \phi_1}{\partial n} \right) ds = 0 \tag{6.95}$$

holds. Application of the far-field forms (6.83) gives

$$c_1^T W c_2 - d_1^T W d_2 = e_1^T W e_2 - f_1^T W f_2 \tag{6.96}$$

and then (6.96) yields the following:

(i) $\phi_1 = \phi_2 = \phi_L \Rightarrow$

$$W R^{(1)^T} = R^{(1)} W, \tag{6.97}$$

(ii) $\phi_1 = \phi_2 = \phi_R \Rightarrow$

$$W R^{(2)^T} = R^{(2)} W, \tag{6.98}$$

(iii) $\phi_1 = \phi_L, \phi_2 = \phi_R \Rightarrow$

$$W T^{(2)^T} = T^{(1)} W, \tag{6.99}$$

(iv) $\phi_1 = \phi_R, \phi_2 = \phi_L \Rightarrow$

$$W T^{(1)^T} = T^{(2)} W. \tag{6.100}$$

Further (6.97)–(6.100) may be used to simplify (6.90)–(6.93) to

$$\boldsymbol{R}^{(1)}\overline{\boldsymbol{R}^{(1)}} + \boldsymbol{T}^{(1)}\overline{\boldsymbol{T}^{(1)}} = \boldsymbol{I}, \tag{6.101}$$

$$\boldsymbol{R}^{(2)}\overline{\boldsymbol{R}^{(2)}} + \boldsymbol{T}^{(2)}\overline{\boldsymbol{T}^{(2)}} = \boldsymbol{I}, \tag{6.102}$$

$$\boldsymbol{R}^{(1)}\overline{\boldsymbol{T}^{(2)}} + \boldsymbol{T}^{(1)}\overline{\boldsymbol{R}^{(2)}} = \boldsymbol{0}, \tag{6.103}$$

$$\boldsymbol{R}^{(2)}\overline{\boldsymbol{T}^{(1)}} + \boldsymbol{T}^{(2)}\overline{\boldsymbol{R}^{(1)}} = \boldsymbol{0}, \tag{6.104}$$

where $\boldsymbol{I}$ is the identity matrix. The above may be written in block matrix form as

$$\begin{pmatrix} \boldsymbol{R}^{(1)} & \boldsymbol{T}^{(1)} \\ \boldsymbol{T}^{(2)} & \boldsymbol{R}^{(2)} \end{pmatrix} \begin{pmatrix} \overline{\boldsymbol{R}^{(1)}} & \overline{\boldsymbol{T}^{(2)}} \\ \overline{\boldsymbol{T}^{(1)}} & \overline{\boldsymbol{R}^{(2)}} \end{pmatrix} = \begin{pmatrix} \boldsymbol{I} & \boldsymbol{0} \\ \boldsymbol{0} & \boldsymbol{I} \end{pmatrix}. \tag{6.105}$$

6.4.3 Multiple gratings

The development of a wide-spacing approximation for diffraction of waves by multiple gratings is similar to that used in §6.3 when there is a single propagating mode. Consider N gratings in the vicinity of $x = L_n$, $n = 1, 2, \dots, N$. For simplicity, it is assumed that the gratings are equally spaced, so that

$$L_{n+1} - L_n \equiv L, \quad n = 1, 2, \dots, N - 1, \tag{6.106}$$

and that for grating n there is a line of symmetry at $x = L_n$. With this assumption the reflection and transmission matrices for grating n satisfy

$$\boldsymbol{R}_n^{(1)} = \boldsymbol{R}_n^{(2)} \equiv \boldsymbol{R}_n \quad \text{and} \quad \boldsymbol{T}_n^{(1)} = \boldsymbol{T}_n^{(2)} \equiv \boldsymbol{T}_n. \tag{6.107}$$

The fluid region is divided up according to the scheme in (6.32). It is convenient to introduce vectors $\boldsymbol{u}_n$ and $\boldsymbol{v}_n$ with components

$$[\boldsymbol{u}_n]_q = e^{ik[(x-L_n)\cos\theta_q + y\sin\theta_q]}, \quad [\boldsymbol{v}_n]_q = e^{ik[-(x-L_n)\cos\theta_q + y\sin\theta_q]}, \tag{6.108}$$

and then

$$\boldsymbol{u} = \boldsymbol{D}_n \boldsymbol{u}_n \quad \text{and} \quad \boldsymbol{v} = \boldsymbol{D}_n^{-1} \boldsymbol{v}_n, \tag{6.109}$$

where $\boldsymbol{D}_n$ is the diagonal matrix with non-zero elements $\{e^{ikL_n\cos\theta_q}, q = -\mu, \dots, \nu\}$ and the components of $\boldsymbol{u}$ and $\boldsymbol{v}$ are given in equation (6.76).

In region n, far enough from the gratings for evanescent waves to be negligible, the potential is written

$$\phi \approx \boldsymbol{a}_n^T \boldsymbol{u}_n + \boldsymbol{b}_n^T \boldsymbol{v}_n, \tag{6.110}$$

where $\boldsymbol{a}_n$ and $\boldsymbol{b}_n$ are unknown at this stage. In region $n+1$, (6.109) is used to express the potential as

$$\phi \approx \boldsymbol{a}_{n+1}^T \boldsymbol{u}_{n+1} + \boldsymbol{b}_{n+1}^T \boldsymbol{v}_{n+1} = \boldsymbol{a}_{n+1}^T \boldsymbol{D}^{-1} \boldsymbol{u}_n + \boldsymbol{b}_{n+1}^T \boldsymbol{D} \boldsymbol{v}_n, \tag{6.111}$$

where $\boldsymbol{D}$ is the diagonal matrix with non-zero elements $\{e^{ikL\cos\theta_q}, \ q = -\mu, \dots, \nu\}$. The waves propagating to the right in region $n+1$ are due to the transmission of the waves propagating to the right in region n and the reflection of the waves propagating to the left in region $n+1$, hence

$$\boldsymbol{a}_{n+1}^T \boldsymbol{D}^{-1} = \boldsymbol{a}_n^T \boldsymbol{T}_n + \boldsymbol{b}_{n+1}^T \boldsymbol{D} \boldsymbol{R}_n. \tag{6.112}$$

Similarly, the waves propagating to the left in region n are due to the transmission of the waves propagating to the left in region $n+1$ and the reflection of the waves propagating to the right in region n, hence

$$\boldsymbol{b}_n^T = \boldsymbol{b}_{n+1}^T \boldsymbol{D} \boldsymbol{T}_n + \boldsymbol{a}_n^T \boldsymbol{R}_n. \tag{6.113}$$

A similar method of solution to that adopted in §6.3 may be followed, but Mulholland and Heckl (1994) suggest a more effective scheme for the calculation of the overall reflection and transmission matrices as follows.

First of all consider the diffraction by two gratings and in particular the two pairs of equations obtained by setting $n = 1, 2$ in (6.112)–(6.113). For a prescribed wave incident from the left $\boldsymbol{a}_1$ is known, $\boldsymbol{b}_3 = \boldsymbol{0}$ and $\boldsymbol{b}_1$ and $\boldsymbol{a}_3$ describe the overall reflection and transmission by the pair of gratings. Elimination of $\boldsymbol{a}_2$ and $\boldsymbol{b}_2$ yields

$$\boldsymbol{b}_1^T = \boldsymbol{a}_1^T \widetilde{\boldsymbol{R}} \quad \text{and} \quad \boldsymbol{a}_3^T = \boldsymbol{a}_1^T \widetilde{\boldsymbol{T}} \boldsymbol{D} \boldsymbol{D}, \tag{6.114}$$

where

$$\widetilde{\boldsymbol{R}} \equiv \boldsymbol{R}_1 + \boldsymbol{T}_1 (\boldsymbol{I} - \boldsymbol{D}\boldsymbol{R}_2\boldsymbol{D}\boldsymbol{R}_1)^{-1} \boldsymbol{D}\boldsymbol{R}_2\boldsymbol{D}\boldsymbol{T}_1 \tag{6.115}$$

and

$$\widetilde{\boldsymbol{T}} \equiv \boldsymbol{T}_1 (\boldsymbol{I} - \boldsymbol{D}\boldsymbol{R}_2\boldsymbol{D}\boldsymbol{R}_1)^{-1} \boldsymbol{D}\boldsymbol{T}_2\boldsymbol{D}^{-1} \tag{6.116}$$

are, respectively, the reflection and transmission matrices for the pair of gratings. The particular form of the transmission matrix $\tilde{T}$ is chosen because

$$a_3^T u_3 = a_3^T D^{-1} D^{-1} u_1 = a_1^T \tilde{T} u_1. \tag{6.117}$$

An iterative algorithm for N gratings can be developed from the observation that if the reflection and transmission matrices are known for n gratings, then these can be regarded as a single grating and the matrices for $n+1$ gratings found from an application of (6.115)-(6.116). The algorithm given by Mulholland and Heckl (1994) is as follows:

(i) $\tilde{R} = R_N$, $\tilde{T} = T_N$;

(ii) for $n = N - 1, 1, -1$ do

$$F = D\tilde{T}D^{-1}, \ G = D\tilde{R}D, \ K = (I - GR_n)^{-1}$$
$$\tilde{R} = R_n + T_n KGT_n, \ \tilde{T} = T_n KF.$$

The algorithm begins by setting the overall reflection and transmission matrices, $\tilde{R}$ and $\tilde{T}$, respectively, to the reflection and transmission matrices for grating N. In the first pass through the loop (ii), the results for two gratings are used to combine the matrices for grating N with those for grating $N - 1$ to obtain updated values for $\tilde{R}$ and $\tilde{T}$. In the next pass through the loop, gratings N and $N - 1$ are regarded as a single entity and combined with grating $N - 2$, again using the results for two gratings. The procedure continues until all N gratings have been accounted for.

Computations for diffraction by multiple gratings are presented in Figure 6.5. The calculations are for normal incidence ($\theta_q = 0$) on gratings of circular cylinders and show how the quantity

$$E_T = \sum_{p=-\mu}^{\nu} |\tilde{T}_{qp}|^2 \cos \theta_p, \tag{6.118}$$

which is proportional to the transmitted wave energy, varies with the number of rows N in the complete grating. For the values of kd chosen there are three propagation directions (see equation 6.72). It is interesting to note the relatively large variations in E_T for $kd = 1.25\pi, 1.5\pi$ compared with $kd = 1.75\pi$ for which E_T varies little with N.

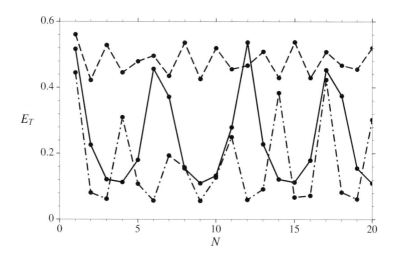

FIGURE 6.5
Transmitted energy E_T for normal wave incidence vs. number of gratings of circular cylinders N; radius $a = 0.5d$, spacing $L = 2d$, $kd = 1.25\pi$ $(- \cdot - \cdot -)$, $kd = 1.5\pi$ (————), $kd = 1.75\pi$ $(- - - -)$.

Bibliographical notes

A detailed investigation of the diffraction of waves by a single grating, that includes calculation methods and reciprocity relations, is given by Twersky (1956, 1962). A multipole method for gratings made up of circular cylinders was devised by Linton and Evans (1993c) and is described here in §3.2.2. The wide-spacing approximation has been applied to the problem of scattering by rows of vertical circular cylinders (that is multiple diffraction gratings) by Dalrymple and Martin (1988). A extensive investigation of multiple gratings has recently been carried out by Heckl (1992), Mulholland and Heckl (1994), and Heckl and Mulholland (1995) in the context of sound propagation through tube bundles. They include an investigation of the relation between diffraction by multiple gratings and propagation of Bloch waves through infinite periodic structures; further results on this topic are given by McIver (2000b).

The idea of using a scattering matrix to develop interaction theories has been used extensively in a great variety of contexts. For example, it has been used to study the scattering of water waves by varying bottom topography (Evans and Linton 1994), carrier transport in semi-conductor devices (Das and Lundstrom 1990), wave propagation in an acoustic waveguide with boundary deformations (Dawson 1991), and elastic wave propagation in multilayered systems (Esquivel-Sirvent and Cocoletzi 1994).

Chapter 7

Wave interaction with small objects

7.1 Introduction

Here techniques are described for obtaining approximations to the solutions of problems in which a wave interacts with an object whose characteristic size is much smaller than the wavelength. The main technique used is the method of matched asymptotic expansions which has been used in a great variety of problems with considerable success. Three problems are chosen here to highlight different applications.

In §7.2 the problem of wave diffraction by a gap in a breakwater, already investigated in §4.7 and §5.1.2, is examined in some detail and used to illuminate some of the main ideas of the method of matched asymptotic expansions. The problem is first solved using an informal technique that does not rely on the application of a formal matching principle. This informal method works well in the gap problem, but is known to give incorrect results in some problems. To avoid this possibility it is advisable to use a formal matching principle and this is described here and then used to solve the gap problem.

The method of matched asymptotic expansions is versatile and can yield results that apply to classes of geometry rather than just to specific geometries. For example, in §7.3 water-wave scattering by a vertical cylinder of uniform, but arbitrary, cross-section is investigated (this problem is treated by the method of integral equations in §4.2.1). The approximate solution obtained here is expressed in terms of the dipole response of the cross-section to a uniform flow which is easily calculated for a number of specific geometries.

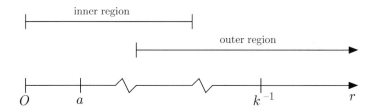

FIGURE 7.1
Sketch of the inner and outer regions.

Both of the problems treated in §§7.2–7.3 have interpretations in the scattering of acoustic and electromagnetic waves as well as water waves. In the latter case, the free surface is removed from the problem in an elementary way. In the third problem of wave radiation by a heaving horizontal cylinder, described in §7.4, this cannot be done and the free surface must be incorporated into the solution procedure.

In §7.5 a fourth problem is considered in which the aim is to find the natural frequencies of oscillation of a fluid in a container that contains a number of structures. Eigenvalue problems of this type can, in principle, be solved by the method of matched asymptotic expansions, but with considerable difficulty. A different technique is described that allows asymptotic results for eigenvalue problems involving small obstacles to be obtained in a quite straightforward way.

The bibliographical notes for this chapter are collected together at the end of the chapter.

7.2 Diffraction by a breakwater

To introduce some of the main ideas of the method of matched asymptotic expansions, the problem of wave scattering by a vertical breakwater with a gap is considered. The breakwater lies along the x-axis with a gap occupying $|x| < a$ and a plane wave with wavenumber k is incident at an angle β to the x-axis. The full details of the problem formulation are given in §4.7 and the geometry is sketched in Figure 4.1. The potential must satisfy the Helmholtz equation and a radiation condition, and have zero normal derivative on the breakwater.

There are two length scales in the problem, namely a and k^{-1} (= wavelength/2π). An approximation to the solution will be obtained under the assumption that the wavelength of the incident waves is significantly greater than the width of the gap so that $ka \ll 1$. To facilitate the solution to the problem the flow domain is thought of as being divided into two overlapping regions. These are an inner region within distances $r = (x^2 + y^2)^{1/2} \ll k^{-1}$ of the aperture and an outer region at distances $r \gg a$ (see Figure 7.1). Separate inner and outer solutions are constructed within each of the regions. Because the full boundary conditions for the problem cannot be applied in both regions, the inner and outer solutions will each contain a number of unknown constants. These constants are determined by matching appropriate asymptotic forms in the overlap region. The inner expansion (as $kr \to 0$) of the outer solution is matched term by term with the outer expansion (as $r/a \to \infty$) of the inner solution according to a well defined 'matching principle'. We begin with an informal approach to the problem in §7.2.1 before describing the formal matching principle in §7.2.2.

7.2.1 Informal solution to the breakwater-gap problem

For low-frequency waves with wavelength significantly greater than the gap width, so that $ka \ll 1$, an observer in the far field will view the effects of the gap as a point disturbance situated on a rigid wall. In particular, the incident wave will drive an oscillatory flow through the gap so that in the far field the disturbance appears as a point wave source at the origin. Thus, the forms of the total potential equivalent to equation (4.112) are the outer solutions

$$\phi_T(x, y) = 2\cos(ky\sin\beta)\, e^{-ikx\cos\beta} + mH_0^{(1)}(kr), \quad y > 0, \qquad (7.1)$$

and

$$\phi_T(x, y) = -mH_0^{(1)}(kr), \quad y < 0, \qquad (7.2)$$

where $H_0^{(1)}(kr)$ is the source solution of the Helmholtz equation and m is the source strength which is to be found. The form for $y < 0$ follows from equation (4.124). The inner expansion of the outer solution in $y > 0$ is

$$\phi_T \sim 2 + m\left\{ 1 + \frac{2i}{\pi}\left(\gamma + \ln\frac{kr}{2}\right)\right\} \quad \text{as} \quad kr \to 0 \qquad (7.3)$$

and in $y < 0$ is

$$\phi_{\mathrm{T}} \sim -m \left\{ 1 + \frac{2\mathrm{i}}{\pi} \left(\gamma + \ln \frac{kr}{2} \right) \right\} \quad \text{as} \quad kr \to 0, \tag{7.4}$$

where the asymptotic form of the Hankel function in (A.17) has been used.

Within the vicinity of the gap the length-scale of the fluid motion is the gap width $2a$ and so, because of the assumption $ka \ll 1$, variations in the fluid motion on the scale of the wavelength cannot be observed and it is appropriate to approximate the field equation by the two-dimensional Laplace equation. The leading-order approximation to the inner solution is therefore constructed from harmonic functions that have zero normal derivative on the surface of the breakwater. Such functions may be determined with the aid of a conformal mapping. The fluid domain in the physical $t = (x + \mathrm{j}y)/a$ plane is mapped onto the upper half of the complex τ plane according to the transformation

$$t = \tfrac{1}{2} \left(\tau + \tau^{-1} \right), \tag{7.5}$$

where the ends of the gap are mapped to $\tau = \pm 1$. In this mapping the square root of minus one is denoted by j to distinguish it from i that arises from the time variation (see equation 4.110). As $|t| \to \infty$ in $\mathrm{Im_j}\, t > 0$,

$$\tau = 2t - \frac{1}{2t} + O\left(t^{-3}\right) \tag{7.6}$$

and as $|t| \to \infty$ in $\mathrm{Im_j}\, t < 0$,

$$\tau = \frac{1}{2t} + O\left(t^{-3}\right). \tag{7.7}$$

Under the mapping, the lines $\mathrm{Im_j}\, \tau = 0$, $\tau \neq 0$, correspond to the breakwater so that the most general solution that satisfies the condition of no flow through the breakwater is

$$\phi_{\mathrm{T}} = Q \ln |\tau| + \sum_{n=-\infty}^{\infty} C_n \, \mathrm{Re_j}\, \tau^n, \tag{7.8}$$

where Q and each C_n are real constants. In the complex τ plane, the origin and the point at infinity in the upper half-plane correspond to the points at infinity in the lower and upper halves of the t plane, hence singularities at these points are permissible in the inner solution. To match with the inner forms of the outer solution in (7.3)–(7.4) which contain constants and logarithms, the general solution is restricted to the form

$$\phi_T = Q \ln |\tau| + C_0 \tag{7.9}$$

which, in view of (7.6)–(7.7) has an outer expansion,

$$\phi_T \sim Q \ln \frac{2r}{a} + C_0 \quad \text{as} \quad \frac{r}{a} \to \infty \tag{7.10}$$

in $y > 0$ and

$$\phi_T \sim -Q \ln \frac{2r}{a} + C_0 \quad \text{as} \quad \frac{r}{a} \to \infty \tag{7.11}$$

in $y < 0$. The inner expansions of the outer solution (7.3)–(7.4) and the outer expansions of the inner solution (7.10)–(7.11) are matched by equating the coefficients of $\ln r$ and the constants, and this results in

$$Q = \frac{2im}{\pi} \quad \text{and} \quad C_0 = 2 + m \left\{ 1 + \frac{2i}{\pi} \left(\gamma + \ln \frac{ka}{4} \right) \right\} \tag{7.12}$$

for the matching in $y > 0$, and

$$Q = \frac{2im}{\pi} \quad \text{and} \quad C_0 = -m \left\{ 1 + \frac{2i}{\pi} \left(\gamma + \ln \frac{ka}{4} \right) \right\} \tag{7.13}$$

for the matching in $y < 0$. Hence

$$m = - \left[1 + \frac{2i}{\pi} \left(\gamma + \ln \frac{ka}{4} \right) \right]^{-1} \tag{7.14}$$

and the outer solution in equations (7.1)–(7.2) is fully determined. From (A.12) and (4.129) the diffraction coefficient $G(\theta, \beta) = 2im$ which is independent of both the angle of incidence β and the angle of observation θ.

7.2.2 A formal matching principle

The above informal approach to the method of matched asymptotic expansions has been used with success in a number of different contexts. However, in some problems erroneous results can be obtained if the method is applied without sufficient care and therefore it is always advisable to apply a formal matching principle. Here only a brief description of a suitable matching principle is given; for further background the reader is recommended to consult Crighton et al. (1992, Chapter 6).

Suppose that a problem depends on the parameter ϵ and inner and outer solutions are to be sought in the limit $\epsilon \to 0$ (in §7.2.1, $\epsilon = ka$). Further, suppose for simplicity that the problem involves a single coordinate r and in the inner region this is scaled to give a non-dimensional coordinate ρ while in the outer region it is scaled to give a non-dimensional coordinate R, where $R = \epsilon\rho$. Note that when ρ is order unity then, by the assumption on ϵ, $R \ll 1$, reflecting the presence of two disparate length scales.

A typical outer solution up to terms in ϵ^m may have the form

$$\chi_0(R) + \epsilon\chi_1(R) + \ldots + \epsilon^m\chi_m(R) \equiv \chi^{(m)}(R;\epsilon), \qquad (7.15)$$

say. In this outer solution put $R = \epsilon\rho$ and expand up to terms in ϵ^n, say, to obtain an inner expansion of the outer solution which is denoted by $\chi^{(m,n)}$. The change of variables is a device to examine the limit $R \to 0$. Similarly, a typical inner solution up to terms in ϵ^n may have the form

$$\psi_0(\rho) + \epsilon\psi_1(\rho) + \ldots + \epsilon^n\psi_n(\rho) \equiv \psi^{(n)}(\rho;\epsilon). \qquad (7.16)$$

In this inner solution put $\rho = R/\epsilon$ and expand up to terms in ϵ^m to obtain an outer expansion of the inner solution which is denoted by $\psi^{(n,m)}$. Now, the change of variables is a device to examine the limit $\rho \to \infty$. The formal matching principle asserts that

$$\chi^{(m,n)} \equiv \psi^{(n,m)}. \qquad (7.17)$$

This is achieved by writing both quantities in terms of the same set of coordinates, either inner or outer, and equating like terms involving the same gauge function. For example, if

$$\chi^{(2,1)} = \epsilon\frac{1}{R} + \epsilon^2\left(\frac{A_1}{R} + \frac{A_2}{R^2}\right) \qquad (7.18)$$

and

$$\psi^{(1,2)} = \epsilon\frac{C_1}{R} + \epsilon^2\left(\frac{\pi}{R} + \frac{\pi}{2R^2}\right) \qquad (7.19)$$

then the matching principle $\chi^{(2,1)} \equiv \psi^{(1,2)}$ requires

$$C_1 = 1, \ A_1 = \pi, \ A_2 = \pi/2. \qquad (7.20)$$

The above contains only integer powers of the small parameter ϵ. In general, m and n may be rational numbers and the series may also contain non-algebraic functions of ϵ. In the following examples gauge functions involving $\ln \epsilon$ appear. When truncating a series in ϵ and applying the matching principle, all terms involving logarithms of ϵ must be grouped with other terms according to the power of ϵ involved. For example, in a wave scattering problem a typical outer solution may have the form

$$
\begin{aligned}
\chi^{(2)} &= \epsilon^2 \left\{ A_0 H_0^{(1)}(R) + A_1 H_1^{(1)}(R) \cos \theta \right\} \\
&= \epsilon^2 \left\{ A_0 H_0^{(1)}(\epsilon\rho) + A_1 H_1^{(1)}(\epsilon\rho) \cos \theta \right\} \\
&= \epsilon^2 \left\{ A_0 \left[1 + \frac{2i}{\pi} \left(\gamma + \ln \frac{\epsilon\rho}{2} \right) + O(\epsilon^2 \ln \epsilon) \right] \right. \\
&\quad \left. + A_1 \left[-\frac{2i}{\pi\epsilon\rho} + O(\epsilon \ln \epsilon) \right] \cos \theta \right\} \quad \text{as} \quad \epsilon \to 0.
\end{aligned}
\tag{7.21}
$$

Hence

$$
\chi^{(2,1)} = \epsilon^2 A_1 \left[-\frac{2i}{\pi\epsilon\rho} \right] \cos \theta
\tag{7.22}
$$

and

$$
\chi^{(2,2)} = \epsilon^2 \left\{ A_0 \left[1 + \frac{2i}{\pi} \left(\gamma + \ln \frac{\epsilon\rho}{2} \right) \right] + A_1 \left[-\frac{2i}{\pi\epsilon\rho} \right] \cos \theta \right\},
\tag{7.23}
$$

where terms in $\epsilon^2 \ln \epsilon$ have been grouped together with terms in ϵ^2. A typical inner solution in such a problem may have an outer expansion

$$
\psi^{(2,2)} = \epsilon^2 \ln \epsilon \, C_0 + \epsilon^2 \left\{ \frac{i \cos \theta}{R} + C_1 + \tfrac{1}{2} \ln \frac{R}{\epsilon} \right\}.
\tag{7.24}
$$

To apply the matching principle $\chi^{(2,2)} \equiv \psi^{(2,2)}$, $\chi^{(2,2)}$ is written in terms of the outer coordinates to get

$$
\chi^{(2,2)} = \epsilon^2 \left\{ A_0 \left[1 + \frac{2i}{\pi} \left(\gamma + \ln \frac{R}{2} \right) \right] + A_1 \left[-\frac{2i}{\pi R} \right] \cos \theta \right\}
\tag{7.25}
$$

(the matching can be carried out equally well with both $\chi^{(2,2)}$ and $\psi^{(2,2)}$ written in terms of the inner coordinates). The application of the matching principle is illustrated in Table 7.1 and results in

$$
A_0 = -\pi i/4, \ A_1 = -\pi/2, \ C_0 = 1/2, \ C_1 = (\gamma - \ln 2 - \pi i/2)/2. \tag{7.26}
$$

gauge function	form	$\chi^{(2,2)}$	$\psi^{(2,2)}$
$\epsilon^2 \ln \epsilon$	constant	0	$C_0 - \frac{1}{2}$
ϵ^2	constant	$A_0 \left[1 + 2\mathrm{i}\left(\gamma - \ln 2\right)/\pi\right]$	C_1
ϵ^2	$\ln R$	$A_0 2\mathrm{i}/\pi$	$\frac{1}{2}$
ϵ^2	$\cos\theta/R$	$A_1\left[-2\mathrm{i}/\pi\right]$	i

Table 7.1   Application of the matching principle $\chi^{(2,2)} \equiv \psi^{(2,2)}$.

7.2.3   Formal solution to the breakwater-gap problem

For the solution of the breakwater-gap problem by a formal application of the method of matched asymptotic expansions it is convenient to decompose the problem into two sub-problems, one symmetric in x and one antisymmetric in x. This is done because the natural gauge functions associated with each problem are different. Thus, the total potential is written

$$\phi_{\mathrm{T}}(x, y) = \phi_{\mathrm{S}}(x, y) + \phi_{\mathrm{A}}(x, y) \qquad (7.27)$$

where

$$\phi_{\mathrm{S}}(-x, y) = \phi_{\mathrm{S}}(x, y) \quad \text{and} \quad \phi_{\mathrm{A}}(-x, y) = -\phi_{\mathrm{A}}(x, y). \qquad (7.28)$$

The symmetric problem

The symmetric problem will be considered first as this is equivalent to the informal solution already obtained in §7.2.1. With the assumption that $\epsilon = ka \ll 1$ and the appearance of source terms in the solution (see equation 7.8), the set of gauge functions in this problem can be expected to involve logarithms of ϵ. However, the informal solution of §7.2.1 gives a diffracted field term that is symmetric in x and for which the denominator contains the factor

$$\ln \frac{ka}{4} + \gamma - \frac{\pi \mathrm{i}}{2} = \ln\left(\frac{-\mathrm{i}ka\,\mathrm{e}^{\gamma}}{4}\right) \equiv \ln\mu \qquad (7.29)$$

say. A formal application of the method of matched asymptotic expansions leads to terms in inverse powers of $\ln\epsilon$ that arise directly from the expansion in ϵ of $[\ln\mu]^{-1}$. The solution procedure can be simplified considerably by adopting $\mu = \sigma\epsilon$ as the expansion parameter, where $\sigma = -\mathrm{i}\,\mathrm{e}^{\gamma}/4$.

The fluid motions in the outer region take place on the scale of the wavelength of the incident waves and hence suitable non-dimensional coordinates are

$$X = kx, \ Y = ky, \ R = kr, \tag{7.30}$$

and it is convenient to solve for the scaled outer potential

$$\chi(X,Y) \equiv \ln\mu\,\phi_{\mathrm{S}}(x,y) \tag{7.31}$$

in order to remove cumbersome logarithmic terms from the denominator (see the informal solution in §7.2.1). Substitution of the scaled outer variables into the governing equations shows that χ satisfies the Helmholtz equation,

$$\frac{\partial^2\chi}{\partial X^2} + \frac{\partial^2\chi}{\partial Y^2} + \chi = 0, \tag{7.32}$$

within the fluid,

$$\frac{\partial\chi}{\partial Y}(X,0) = 0, \quad |X| > \epsilon, \tag{7.33}$$

and a radiation condition to ensure that the diffracted part of the field propagates away from the gap. In the limit $\epsilon \to 0$, (7.33) becomes

$$\frac{\partial\chi}{\partial Y}(X,0) = 0, \quad |X| \neq 0, \tag{7.34}$$

and it is this boundary condition that is to be applied in the outer region for the solution by matched asymptotic expansions.

Guided by equations (7.1) and (7.2), the leading-order outer solution is written

$$\chi_+^{(0)} = \ln\mu.2\cos(X\cos\beta)\cos(Y\sin\beta) + A_0 H_0^{(1)}(R), \ Y > 0, \tag{7.35}$$

and

$$\chi_-^{(0)} = -A_0 H_0^{(1)}(R), \ Y < 0, \tag{7.36}$$

where A_0 is a constant to be determined from the matching. The first term on the right-hand side of (7.35) is the symmetric part of the incident wave. The most general form for the diffracted field in $Y > 0$ that is consistent with the boundary and symmetry conditions is

$$\sum_{n=0}^{\infty} A_{2n} H_{2n}^{(1)}(R) \cos 2n\theta,$$

but the subsequent matching gives $A_{2n} = 0$, $n > 0$, so for simplicity of presentation only the first term in the series is included. In general, the user of the method should be aware of the most general solution at each stage and include appropriate additional terms if difficulties arise in the matching.

The variations in the fluid motion in the inner region take place on the scale of the width of the breakwater gap and so suitable non-dimensional coordinates are

$$\xi = \frac{x}{a}, \ \eta = \frac{y}{a}, \ \rho = \frac{r}{a}. \tag{7.37}$$

When the outer solutions $\chi_{\pm}^{(0)}$ are written in terms of the inner coordinates and expanded up to terms in $\ln \mu$ and constants, then the inner expansions of the outer solutions are

$$\chi_{+}^{(0,0)} = 2 \ln \mu + A_0 \left\{ 1 + \frac{2i}{\pi} \left(\gamma + \ln \frac{\mu\rho}{2\sigma} \right) \right\} = 2 \ln \mu + A_0 \frac{2i}{\pi} \ln 2\mu\rho \tag{7.38}$$

and

$$\chi_{-}^{(0,0)} = -A_0 \left\{ 1 + \frac{2i}{\pi} \left(\gamma + \ln \frac{\mu\rho}{2\sigma} \right) \right\} = -A_0 \frac{2i}{\pi} \ln 2\mu\rho, \tag{7.39}$$

where the definition of σ has been used to simplify the expressions.

When rewritten in terms of the inner coordinates (7.37), the governing equations for the the inner potential $\psi(\xi, \eta) \equiv \phi_S(x, y)$ are

$$\frac{\partial^2 \psi}{\partial \xi^2} + \frac{\partial^2 \psi}{\partial \eta^2} + \frac{\mu^2}{\sigma^2} \psi = 0 \tag{7.40}$$

within the fluid and

$$\frac{\partial \psi}{\partial \eta}(\xi, 0) = 0, \quad |\xi| > 1. \tag{7.41}$$

The inner solution turns out to have the form

$$\psi(\xi, \eta; \mu) = \ln \mu \ \psi_{01}(\xi, \eta) + \psi_0(\xi, \eta) + \ldots + \mu^2 \psi_2(\xi, \eta) + \ldots . \tag{7.42}$$

When this expansion is substituted into (7.40) and terms multiplied by the same function of μ equated, it may be seen that all terms up to, but **not** including, that in μ^2 are harmonic functions (that is, solutions of the two-dimensional Laplace equation). Terms from that in μ^2 onwards will satisfy a Poisson equation if there is a non-zero term at an order μ^2 lower in the expansion. For example, ψ_2 is a solution of

$$\frac{\partial^2 \psi_2}{\partial \xi^2} + \frac{\partial^2 \psi_2}{\partial \eta^2} = -\frac{\psi_0}{\sigma^2}. \tag{7.43}$$

In what follows only the expansion $\psi^{(0)}$ (that is, all terms in ψ up to and including ϵ^0) is needed and so all the functions that appear are harmonic.

The radiation condition is not relevant to the inner region and so the leading-order approximation to the inner potential is a harmonic function that satisfies (7.41). Such functions have already been discussed in §7.2.1 and may be expressed in terms of the complex variable τ introduced in equation (7.5), where now $t = \xi + j\eta$. The general form of the inner solution is given in (7.8). However, in view of the requirement to match with the forms in (7.38)–(7.39), the leading-order inner solution is written

$$\psi^{(0)} = \ln \mu . B_0 + Q \ln |\tau| + C_0, \tag{7.44}$$

where B_0, Q and C_0 are constants to be found from the matching. When the inner solution $\psi^{(0)}$ is, with the aid of (7.6)–(7.7), written in terms of the outer coordinates, expanded up to terms in $\ln \mu$ and constants and then rewritten back in inner coordinates it is found that

$$\psi_\pm^{(0,0)} = \ln \mu . B_0 \pm Q \ln 2\rho + C_0, \tag{7.45}$$

where the upper sign indicates the expansion in $\eta > 0$ and the lower sign the expansion in $\eta < 0$. Application of the matching principles $\psi_\pm^{(0,0)} \equiv \chi_\pm^{(0,0)}$ yields

$$C_0 = 0, \ B_0 = 1, \ A_0 = \pi i/2, \ Q = -1, \tag{7.46}$$

so that

$$\psi^{(0)} = \ln \mu - \ln |\tau|. \tag{7.47}$$

Further expansion of the inner solution, again using (7.6)–(7.7), gives

$$\psi_\pm^{(0,2)} = \ln \mu \mp \ln 2\rho \pm \frac{\cos 2\theta}{4\rho^2} \tag{7.48}$$

which, in view of the relation $\rho = R/\epsilon = \sigma R/\mu$ between the inner and outer coordinates, suggests the need for a dipole in the outer solution at order μ^2. Hence the outer solutions are extended to

$$\chi_+^{(2)} = \ln \mu.2 \cos(X \cos \beta) \cos(Y \sin \beta)$$
$$+ \frac{\pi i}{2} H_0^{(1)}(R) + \mu^2 A_2 H_2^{(1)}(R) \cos 2\theta \quad (7.49)$$

and

$$\chi_-^{(2)} = -\frac{\pi i}{2} H_0^{(1)}(R) - \mu^2 A_2 H_2^{(1)}(R) \cos 2\theta \quad (7.50)$$

which have inner expansions

$$\chi_\pm^{(2,0)} = \ln \mu \mp \ln 2\rho \mp \frac{4i\sigma^2 A_2}{\pi} \frac{\cos 2\theta}{\rho^2} \quad (7.51)$$

and the matching principles $\psi_\pm^{(0,2)} \equiv \chi_\pm^{(2,0)}$ yield

$$A_2 = \pi i/16\sigma^2. \quad (7.52)$$

This information is enough to determine the symmetric outer potential up to terms in μ^2 and, in terms of the original variables, this gives

$$\phi_S^{(2)} = 2 \cos(kx \cos \beta) \cos(ky \sin \beta) + \frac{\pi i}{2 \ln \mu} H_0^{(1)}(kr)$$
$$+ \frac{\pi i \epsilon^2}{16 \ln \mu} H_2^{(1)}(kr) \cos 2\theta, \ y > 0. \quad (7.53)$$

Note that to this order of approximation the symmetric part of the diffracted field in the outer region is independent of the angle of incidence β.

The antisymmetric problem

For the antisymmetric problem expansions are sought in terms of the parameter $\epsilon = ka \ll 1$ as there is no information to suggest another parameter is more appropriate. The scalings adopted are the same as for the symmetric problem (see equations 7.30 and 7.37) and in order to satisfy (7.32)–(7.33) the outer solution is written

$$\chi_+^{(p)} = -2i \sin(X \cos \beta) \cos(Y \sin \beta) + f_p(\epsilon) A_p H_1^{(1)}(R) \cos \theta \quad (7.54)$$

for $Y > 0$ and

$$\chi_-^{(p)} = -f_p(\epsilon) A_p H_1^{(1)}(R) \cos \theta \qquad (7.55)$$

for $Y < 0$. The first term on the right-hand side of (7.54) is the anti-symmetric part of the incident wave. The second term contains the least-singular term that satisfies the required asymmetry and has zero normal derivative on $Y = 0$, $X \neq 0$. If this is an incorrect choice then this would be revealed in the failure of the matching. The size of this term is not known at this stage, hence the presence of the unknown gauge function $f_p(\epsilon)$; this and the constant A_p will be determined by the matching.

When the outer solutions are rewritten in terms of the inner variables and expanded using (7.6)–(7.7) then, from equations (A.16) and (A.18), as $\epsilon \to 0$

$$\chi_+^{(p)} = -2i\epsilon\xi \cos \beta + O(\epsilon^3) + f_p(\epsilon) A_p \left(-\frac{2i \cos \theta}{\pi \epsilon \rho} + O(\epsilon) \right) \qquad (7.56)$$

and

$$\chi_-^{(p)} = -f_p(\epsilon) A_p \left(-\frac{2i \cos \theta}{\pi \epsilon \rho} + O(\epsilon) \right). \qquad (7.57)$$

A general form for the leading-order inner solution is

$$\psi^{(q)} = g_q(\epsilon) \sum_{\substack{n=-\infty \\ |n| \text{ odd}}}^{\infty} C_n \operatorname{Re_j} \tau^n, \qquad (7.58)$$

where the gauge function $g_q(\epsilon)$ is to be found (the selection of the odd integers arises from the requirement that the solution be antisymmetric in x). Suppose that the non-zero term with the largest exponent involves τ^m, where m is a positive integer. In the upper half-plane $\tau \sim 2t$ as $|t| \to \infty$ so that a term proportional to R^m would then be forced at leading order in the inner expansion (7.56) of the outer solution for $Y > 0$. The only term of this type is with $m = 1$. Now suppose that the term with smallest exponent involves τ^{-m}, where m is a positive integer. In the lower half-plane $\tau \sim 1/2t$ as $|t| \to \infty$ so that a term proportional to R^m would then forced at leading order in the inner expansion (7.57) of the outer solution for $Y < 0$. There is no term of this type. Hence, the only possibility of matching with the outer solution is if the leading-order inner solution has the form

$$\psi^{(q)} = g_q(\epsilon) C_0 \operatorname{Re_j} \tau \qquad (7.59)$$

and hence

$$\psi_+^{(q,q-1)} = g_q(\epsilon) C_0 2\xi. \tag{7.60}$$

From (7.56) it is now apparent that the matching will work only if $q = 1$, $g_1(\epsilon) = \epsilon$ and $f_p(\epsilon) = o(1)$ as $\epsilon \to 0$, and hence

$$\chi_+^{(0,1)} = -2i\epsilon\xi \cos\beta \tag{7.61}$$

so that the matching principle $\chi_+^{(0,1)} \equiv \psi_+^{(1,0)}$ gives

$$C_0 = -i \cos\beta. \tag{7.62}$$

Further expansion of the inner solution gives

$$\psi_+^{(1,2)} = -i\epsilon \cos\beta \left(2\xi - \frac{\cos\theta}{2\rho} \right) \tag{7.63}$$

and matching with (7.56) is possible only if $p = 2$ and $f_2(\epsilon) = \epsilon^2$ so that the matching principle $\chi_+^{(2,1)} = \psi_+^{(1,2)}$ gives

$$A_2 = -\tfrac{1}{4}\pi \cos\beta. \tag{7.64}$$

The antisymmetric potential up to terms in ϵ^2 is now determined and is

$$\phi_{\mathrm{A}}^{(2)} = -2i \sin(kx \cos\beta) \cos(ky \sin\beta) - \tfrac{1}{4}\epsilon^2\pi \cos\beta\, H_1^{(1)}(kr) \cos\theta. \tag{7.65}$$

The diffraction coefficient

From (A.12), (7.53), and (7.65) the low-frequency approximation to the diffraction coefficient defined in (4.129) is

$$G(\theta, \beta) = \pi \left(-\frac{1}{\ln\mu} + \frac{(ka)^2}{8\ln\mu} \cos 2\theta - \tfrac{1}{2}(ka)^2 \cos\beta \cos\theta \right), \tag{7.66}$$

where μ is defined in equation (7.29). This low-frequency approximation is compared with numerical solutions in Figure 7.2 for $ka = \pi/8$. This comparison is typical for $ka \in (0, \pi/4)$ in that the approximation captures the essential features of the exact solution without achieving high numerical accuracy. The maximum numerical error in $|G(\theta, \beta)|$ ranges from about 0.1% at $ka = \pi/32$ to about 20% at $ka = \pi/4$. By $ka = \pi/2$, the low-frequency approximation has broken down and does not reproduce the oscillations of the exact solution as θ varies.

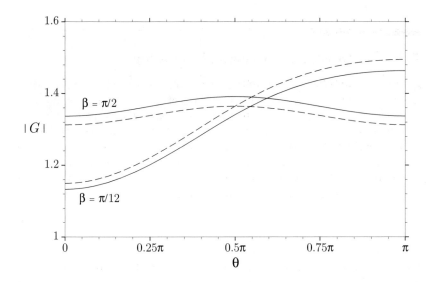

FIGURE 7.2
The modulus of the diffraction coefficient $G(\theta, \beta)$ for a break-water with a gap for $ka = \pi/8$. Accurate numerical results are shown with a solid line and the low-frequency approximation with a dashed line.

7.2.4 The insular breakwater

It is relatively straightforward to find approximations to the insular breakwater problem by the method of matched asymptotic expansions (it is a special case of the problem treated later in §7.3). However, the result (4.164) allows an approximation to the diffraction coefficient to be deduced immediately from the approximation for the breakwater-gap problem given in equation (7.66). The result is

$$F(\theta, \beta) = \tfrac{1}{2}\pi(ka)^2 \sin\theta \sin\beta. \qquad (7.67)$$

In this approximation the far-field diffracted waves are always symmetric about $\theta = \pi/2$, which is not true for the exact solution.

7.3   Scattering by a vertical cylinder

Here the problem of scattering of a plane wave by a cylinder with arbitrary cross-section is solved using the method of matched asymptotic expansions and used to illustrate that the precise geometry need not be specified in order to apply the method. An integral equation formulation of this problem is given in §4.2.1. The cylinder is assumed to have uniform cross-section throughout the depth so that the depth can be removed from the problem. A plane wave of wavenumber k propagates in the direction of x increasing and is taken to have potential

$$\phi_{\mathrm{I}} = \mathrm{e}^{-\mathrm{i}kr\cos(\theta-\beta)} . \tag{7.68}$$

The boundary of the cross-section of the cylinder is denoted by Γ and the origin of the horizontal coordinates (x, y) is chosen to be within Γ. The potential ϕ for the scattered wave field must satisfy the Helmholtz equation

$$(\nabla^2 + k^2)\phi = 0 \tag{7.69}$$

within the fluid region, the boundary condition

$$\frac{\partial\phi}{\partial n} = -\frac{\partial\phi_{\mathrm{I}}}{\partial n} \quad \text{on} \quad \Gamma, \tag{7.70}$$

where n is a normal coordinate directed out of the fluid, and a radiation condition in the form (1.29) that specifies outgoing waves.

An approximation to the solution will be obtained under the assumption that $\epsilon = ka \ll 1$, where a is a measure of the horizontal dimension of Γ. The fluid domain is divided into an inner region surrounding the cylinder to distances $r = (x^2 + y^2)^{1/2} \ll k^{-1}$ and an outer region at distances $r \gg a$. Scaled coordinates for the outer region are defined by

$$X = kx, \; Y = ky, \; R = kr, \tag{7.71}$$

so that the field equation for the outer potential $\chi(X, Y) \equiv \phi(x, y)$ is

$$\frac{\partial^2\chi}{\partial X^2} + \frac{\partial^2\chi}{\partial Y^2} + \chi = 0. \tag{7.72}$$

The boundary condition (7.70) is not relevant to the outer region but χ must satisfy the radiation condition.

In the inner region, scaled coordinates are defined by

$$\xi = \frac{x}{a}, \quad \eta = \frac{y}{a}, \quad \rho = \frac{r}{a}, \tag{7.73}$$

and in terms of these variables the field equation for the inner potential $\psi(\xi, \eta) \equiv \phi(x, y)$ becomes

$$\frac{\partial^2 \psi}{\partial \xi^2} + \frac{\partial^2 \psi}{\partial \eta^2} + \epsilon^2 \psi = 0 \tag{7.74}$$

and the boundary condition (7.70) becomes

$$\frac{\partial \psi}{\partial n} = -\frac{\partial}{\partial n}\left[e^{-i\epsilon\rho\cos(\theta-\beta)}\right] = i\epsilon\frac{\partial}{\partial n}\left[\rho\cos(\theta - \beta)\right]$$

$$+ \tfrac{1}{2}\epsilon^2\frac{\partial}{\partial n}\left[\rho^2\cos^2(\theta - \beta)\right] + \dots \quad \text{on} \quad \Gamma. \tag{7.75}$$

This indicates the inner solution has terms at orders ϵ and ϵ^2. The possibility that an expansion contains other terms should always be borne in mind and the matching will fail if required terms are omitted. Thus, an inner solution might be examined in the form

$$\psi^{(1)} = g_p(\epsilon)\psi_p + \epsilon\psi_1 + g_q(\epsilon)\psi_q + \epsilon^2\psi_2, \tag{7.76}$$

where as $\epsilon \to 0$, $g_p(\epsilon) = o(\epsilon)$, $\epsilon = o(g_q(\epsilon))$, and $g_q(\epsilon) = o(\epsilon^2)$, and the matching used to determine the gauge functions $g_p(\epsilon)$ and $g_q(\epsilon)$ as was done for the antisymmetric problem in §7.2.3. Here, this process gives an inner expansion in the form

$$\psi^{(1)} = \epsilon\psi_1 + \epsilon^2\ln\epsilon\,\psi_{21} + \epsilon^2\psi_2, \tag{7.77}$$

where after substitution in (7.74) and (7.75) and comparison of like terms, ψ_1, ψ_{21}, and ψ_2 are seen to be harmonic functions that satisfy

$$\frac{\partial\psi_1}{\partial n} = i\frac{\partial}{\partial n}\left[\rho\cos(\theta - \beta)\right] \quad \text{on} \quad \Gamma, \tag{7.78}$$

$$\frac{\partial\psi_{21}}{\partial n} = 0 \quad \text{on} \quad \Gamma, \tag{7.79}$$

and

$$\frac{\partial\psi_2}{\partial n} = \tfrac{1}{2}\frac{\partial}{\partial n}\left[\rho^2\cos^2(\theta - \beta)\right] \quad \text{on} \quad \Gamma, \tag{7.80}$$

respectively. The boundary condition (7.78) is the same as that in the problem of a uniform stream of complex velocity i flowing past the cylinder from the direction $\theta = \beta$ and hence ψ_1 is just the disturbance to this flow generated by the cylinder. From results in potential theory (Batchelor 1967, equation 2.10.4)

$$\psi_1 \sim \mathrm{i} \left(\frac{\mu_x \cos\theta}{\rho} + \frac{\mu_y \sin\theta}{\rho} \right) \quad \text{as} \quad \rho \to \infty, \tag{7.81}$$

where a possible constant term has been omitted as this cannot be matched with the outer solution given below. The dipole coefficients μ_x and μ_y are properties of the geometry. The only solution for ψ_{12} consistent with the matching is a constant C_0, say. From the divergence theorem

$$\int_\Gamma \frac{\partial \psi_2}{\partial n} \, \mathrm{d}s = \int_\Gamma \frac{1}{2} \frac{\partial}{\partial n} \left[\rho^2 \cos^2(\theta - \beta) \right] \mathrm{d}s$$
$$= -\frac{1}{2} \int_S \nabla_{\xi\eta}^2 \left[(\xi \cos\beta + \eta \sin\beta)^2 \right] \mathrm{d}S = -S, \tag{7.82}$$

where S is the area within Γ, and hence (Batchelor 1967, equation 2.10.4)

$$\psi_2 \sim C_1 + \frac{S}{2\pi} \ln\rho \quad \text{as} \quad \rho \to \infty, \tag{7.83}$$

where C_1 is a constant (note that Batchelor takes the normal to be directed into the fluid). From the above results

$$\psi^{(2,2)} = \epsilon^2 \ln\epsilon \, C_0$$
$$+ \epsilon^2 \left\{ \mathrm{i} \left(\frac{\mu_x \cos\theta}{R} + \frac{\mu_y \sin\theta}{R} \right) + C_1 + \frac{S}{2\pi} \ln\frac{R}{\epsilon} \right\}. \tag{7.84}$$

The outer expansion of the inner solution (7.84) suggests that the leading-order outer solution has the form

$$\chi^{(2)} = \epsilon^2 \left\{ A_0 H_0^{(1)}(R) + A_1 H_1^{(1)}(R) \cos\theta + B_1 H_1^{(1)}(R) \sin\theta \right\}, \tag{7.85}$$

where A_0, A_1, and B_1 are constants to be found from the matching, and this has an inner expansion

$$\chi^{(2,2)} = \epsilon^2 \left\{ A_0 \left[1 + \frac{2\mathrm{i}}{\pi} \left(\gamma + \ln\frac{R}{2} \right) \right] \right.$$
$$\left. + (A_1 \cos\theta + B_1 \sin\theta) \left(-\frac{2\mathrm{i}}{\pi R} \right) \right\}. \tag{7.86}$$

The matching principle $\chi^{(2,2)} = \psi^{(2,2)}$ yields

$$A_0 = -\tfrac{1}{4}iS, \quad A_1 = -\tfrac{1}{2}\pi\mu_x, \quad B_1 = -\tfrac{1}{2}\pi\mu_y, \tag{7.87}$$

$$C_0 = \frac{S}{2\pi}, \quad C_1 = \frac{S}{2\pi}\left(\gamma - \ln 2 - \frac{\pi i}{2}\right) \tag{7.88}$$

and hence the outer solution $\chi^{(2)}$ is fully determined

The dipole coefficients for a number of cylinder contours Γ are readily available. For example, Milne-Thomson (1996, p. 167) solves the problem of flow past an elliptical cylinder with axes of length $2a$ and $2b$ measured parallel to the ξ- and η-axes, and the dipole coefficients are readily calculated as

$$\mu_x = -\frac{(a+b)b\cos\beta}{2a^2} \quad \text{and} \quad \mu_y = -\frac{(a+b)\sin\beta}{2a}. \tag{7.89}$$

For the particular case of a flat plate of length $2a$ lying along the x-axis, so that $b = S = 0$,

$$\chi^{(2)} = \tfrac{1}{4}\epsilon^2\pi\sin\beta \; H_1^{(1)}(R)\sin\theta$$

$$\sim \left(\frac{e^{i(R-3\pi/4)}}{2\pi R}\right)^{1/2} \tfrac{1}{2}\epsilon^2\pi\sin\beta\sin\theta \quad \text{as} \quad R \to \infty, \tag{7.90}$$

which confirms the result in equation (7.67).

7.4   Radiation by a heaving cylinder

Water-wave problems involving a free surface may also be solved by the method of matched asymptotic expansions. To illustrate the main ideas, the two-dimensional problem of wave radiation due to the vertical (heave) oscillations of a half-immersed, horizontal, circular cylinder of radius a in deep water is considered. The origin of coordinates is chosen to be at the centre of the cylinder so that the mean free surface is $z = 0$, $|x| > a$. The fluid motion takes place in the x, z-plane and polar coordinates (r, θ) are defined by

$$x = r\sin\theta, \quad z = -r\cos\theta \tag{7.91}$$

so that θ is measured from the downward vertical as in §3.1.1.

The problem is formulated as in §1.3.3 so that the time-independent velocity potential $\phi_3(x, z)$ for the flow satisfies the Laplace equation within the fluid, the free surface condition (1.13), the decay condition (1.15), the cylinder boundary condition

$$\frac{\partial \phi_3}{\partial r} = -\cos\theta \quad \text{on} \quad r = a, \ |\theta| < \pi/2 \qquad (7.92)$$

(see equation 1.39), and a radiation condition specifying that the waves generated are outgoing. There are two length scales in the problem, namely the cylinder radius a and the wavelength of the radiated waves $2\pi/K$, where K appears in the free-surface condition (1.13). An approximate solution is sought under the assumption that $\epsilon = Ka \ll 1$. The inner region is the fluid region surrounding the cylinder to distances $r \ll K^{-1}$ and the outer region is the fluid in $r \gg a$.

Define inner region coordinates by

$$\xi = \frac{x}{a}, \ \zeta = \frac{z}{a}, \ \rho = \frac{r}{a}. \qquad (7.93)$$

With this change of variables, the cylinder and free-surface boundary conditions for the inner potential $\psi(\xi, \zeta) \equiv \phi_3(x, z)/a$ become, respectively,

$$\frac{\partial \psi}{\partial \rho} = -\cos\theta \quad \text{on} \quad \rho = 1, \ |\theta| < \pi/2, \qquad (7.94)$$

and

$$\frac{\partial \psi}{\partial \zeta} = \epsilon\psi \quad \text{on} \quad \zeta = 0, \ |\xi| > 1. \qquad (7.95)$$

As $\epsilon \to 0$, the leading-order approximation to the free-surface boundary condition is

$$\frac{\partial \psi}{\partial \theta} = 0 \quad \text{on} \quad |\theta| = \pi/2, \ \rho > a. \qquad (7.96)$$

A harmonic function satisfying (7.94) and (7.96) is

$$\psi_{\text{p}} = -\frac{2}{\pi}\left(\ln\rho + \sum_{n=1}^{\infty} \frac{(-1)^n \cos 2n\theta}{n(4n^2 - 1)\rho^{2n}}\right). \qquad (7.97)$$

Solutions satisfying homogeneous boundary conditions could be added to this, but apart from constants such solutions grow as $\rho \to \infty$ and prove impossible to match with the outer solution. The outer expansion of ψ_p contains logarithms and the previous examples suggest that this requires constant terms at order $\ln \epsilon$ in the inner solution. Thus, the leading-order inner solution is written

$$\psi^{(0)} = \ln \epsilon \, C_0 + C_1 + \psi_p \tag{7.98}$$

which has an outer expansion

$$\psi^{(0,0)} = \ln \epsilon \, C_0 + C_1 - \frac{2}{\pi} \ln \frac{R}{\epsilon}. \tag{7.99}$$

Define outer region coordinates by

$$X = Kx, \ Z = Kz, \ R = Kr \tag{7.100}$$

and the outer potential $\chi(X, Z) \equiv \phi_3(x, z)$. In the outer region the cylinder appears as a point disturbance and the free surface condition is to be applied on $y = 0$, $x \neq 0$. The outer expansion of the inner solution suggests that the leading-order outer solution is a wave source at the origin so that

$$\chi^{(0)} = A_0 g_0(X, Z), \tag{7.101}$$

where

$$g_0(X, Z) = -\int_0^\infty \frac{e^{\mu Z}}{\mu - 1} \cos \mu X \, d\mu \tag{7.102}$$

is given in equation (B.13). From (B.16) the inner expansion is

$$\chi^{(0,0)} = A_0 \left(\gamma + \ln R - \pi i \right) \tag{7.103}$$

so that the matching principle $\chi^{(0,0)} \equiv \psi^{(0,0)}$ gives

$$A_0 = -2/\pi, \ C_0 = -2/\pi, \ C_1 = -2(\gamma - \pi i)/\pi. \tag{7.104}$$

Asymptotic approximations as $Ka \to 0$ to the added mass and damping coefficients per unit length of the cylinder follow from equation (1.48) and are given by

$$a_{33} + \frac{ib_{33}}{\omega} \sim \rho \int_{-\pi/2}^{\pi/2} [a\psi^{(0)}] \cos \theta \, a \, d\theta$$

$$= \rho a^2 \int_{-\pi/2}^{\pi/2} \left[-\frac{2}{\pi} (\ln Ka + \gamma - \pi i) + \psi_p \right] \cos \theta \, d\theta. \tag{7.105}$$

The series that results from the integration of ψ_p can be summed (Gradshteyn and Ryzhik 1980, equation 0.236.5) and thus as $Ka \to 0$

$$a_{33} + \frac{ib_{33}}{\omega} \sim \frac{4\rho a^2}{\pi}\left(-\ln Ka + \tfrac{3}{2} - 2\ln 2 - \gamma + \pi i\right). \qquad (7.106)$$

7.5 A technique for eigenvalue problems

An interesting technique for the calculation of the natural frequencies of oscillation of a fluid within a container that has a number of small objects within it is given by Curzon and Plant (1986). Here the method is described in the context of a sloshing problem considered by Drake (1999). The concrete shaft of the Draugen oil-production platform contains a number of pipes and the aim is to calculate the effect of these pipes on the fundamental sloshing frequency of water contained within the shaft. The concrete shaft and the N pipes within the shaft are all simplified to rigid vertical circular cylinders; a cross-section through the geometry is illustrated in Figure 7.3. Within this cross-section, D denotes the the fluid domain (excluding the pipes) and S_0 the outer boundary. The cross-sectional area of pipe i is denoted by D_i and the corresponding boundary by S_i. Cartesian coordinates (x, y, z) are chosen so that the x, y-plane is parallel to the cross-section and the z-axis is directed vertically upwards.

As in §2.3.5, the velocity potential is written $\phi(x, y, z) = \varphi(x, y)\psi_0(z)$, where $\psi_0(z)$ is the vertical eigenfunction defined in equation (2.11), and hence

$$-\nabla^2\varphi = k^2\varphi \quad \text{for} \quad (x, y) \in D \qquad (7.107)$$

and

$$\frac{\partial\varphi}{\partial n} = 0 \quad \text{for} \quad (x, y) \in S_i, \quad i = 0, 1, \ldots N. \qquad (7.108)$$

The aim is determine the eigenvalues of the Laplacian k^2 and hence, through equation (2.9), the frequencies of oscillation of the fluid.

Let k_0^2 be an eigenvalue of the Laplacian for the domain D_0 bounded by S_0 (that is with the pipes removed) when a homogeneous Neumann condition is imposed on S_0, and let φ_0 be the corresponding eigenfunc-

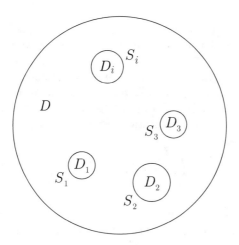

FIGURE 7.3
**Cross-section of the shaft geometry. (Adapted from Figure 9
of K. R. Drake, *Applied Ocean Research*, 21, 133–143, 1999.)**

tion. Thus,

$$-\nabla^2\varphi_0 = k_0^2\varphi_0 \quad \text{for} \quad (x,y) \in D_0 \tag{7.109}$$

and

$$\frac{\partial\varphi_0}{\partial n} = 0 \quad \text{for} \quad (x,y) \in S_0. \tag{7.110}$$

An application of Green's theorem (see §4.2) to φ and φ_0 over their
common domain of definition D gives

$$\iint_D \left(\varphi_0\nabla^2\varphi - \varphi\nabla^2\varphi_0\right)\,\mathrm{d}A = \int_\Gamma \left(\varphi_0\frac{\partial\varphi}{\partial n} - \varphi\frac{\partial\varphi_0}{\partial n}\right)\,\mathrm{d}s, \tag{7.111}$$

where $\Gamma = \cup_{i=0}^n S_i$ is the boundary of D. With the aid of equations
(7.107)–(7.110) this simplifies to

$$(k_0^2 - k^2)\iint_D \varphi_0\varphi\,\mathrm{d}A = -\int_S \varphi\frac{\partial\varphi_0}{\partial n}\,\mathrm{d}s, \tag{7.112}$$

where $S = \cup_{i=1}^n S_i$ is the boundary of the pipes. Under the assumption
that each pipe S_i has small radius a_i relative to the primary length scale

of the fluid motion (that is $ka_i \ll 1$, $i = 1, 2, \ldots, N$), an approximation to the eigenvalue k^2 is obtained by approximating the integrals in (7.112).

For pipes of sufficiently small radius the function φ_0 is a good approximation to the eigenfunction φ over the bulk of the domain D and furthermore, in the integral on the left-hand side of (7.112), φ may be replaced by φ_0 and the integration carried out over the whole interior D_0 of S_0 so that

$$\iint_D \varphi_0 \varphi \, dA \approx \iint_{D_0} \varphi_0^2 \, dA. \tag{7.113}$$

In the boundary integrations over $\{S_i; \; i = 1, 2, \ldots, N\}$, the perturbation of φ from φ_0 is significant because on each S_i the normal velocity must be zero. In the vicinity of each S_i write

$$\varphi = \varphi_0 + \varphi_i \tag{7.114}$$

where, because of the assumption $ka_i \ll 1$, φ_i is locally a solution of Laplace's equation (see equation 7.74, for example). By Taylor expansion about the centre O_i of S_i, the potential φ_0 at a point with position vector $\boldsymbol{r}$ is, in the limit $a_i \to 0$,

$$\varphi_0(\boldsymbol{r}) = \varphi_{0i} + \nabla_i \varphi_0 . \boldsymbol{R}_i + O(a_i^2) \tag{7.115}$$

where $\boldsymbol{R}_i$ is the position vector relative to O_i, and φ_{0i} and $\nabla_i \varphi_0$ are, respectively, the values of φ_0 and $\nabla \varphi_0$ at O_i. The boundary condition of no flow through S_i is therefore

$$\frac{\partial \varphi_i}{\partial R_i} = -\frac{\partial \varphi_0}{\partial R_i} \approx -\frac{\nabla_i \varphi_0 . \boldsymbol{R}_i}{R_i} \quad \text{on} \quad R_i = a_i, \tag{7.116}$$

provided a_i is sufficiently small, and therefore

$$\varphi_i \approx \frac{a_i^2 \nabla_i \varphi_0 . \boldsymbol{R}_i}{R_i^2} \tag{7.117}$$

after requiring that the disturbance $\varphi_i \to 0$ as $R_i/a_i \to \infty$. The approximation (7.117) to the perturbation potential φ_i is the dipole response to the near uniform flow generated by φ_0 in the neighbourhood of S_i (see equation 7.81). With this approximation to φ_i, the right-hand side of (7.112)

$$\int_S \varphi \frac{\partial \varphi_0}{\partial n} \, ds \approx -\sum_{i=1}^n \int_{S_i} (\varphi_0 + \nabla_i \varphi_0 . \boldsymbol{R}_i) \frac{\partial \varphi_0}{\partial R_i} \, ds, \tag{7.118}$$

where the minus sign arises because the normal coordinate n is directed out of the fluid region. Application of the divergence theorem and the assumption that the variation of φ_0 and $\nabla\varphi_0$ across D_i may be neglected gives

$$\int_{S_i} \varphi_0 \frac{\partial\varphi_0}{\partial R_i}\, \mathrm{d}S = \iint_{D_i} \nabla.(\varphi_0\nabla\varphi_0)\, \mathrm{d}A = \iint_{D_i} \left(\varphi_0\nabla^2\varphi_0 + |\nabla\varphi_0|^2\right)\, \mathrm{d}A$$

$$\approx \pi a_i^2 \left(\varphi_{0i}\nabla_i^2\varphi_0 + |\nabla_i\varphi_0|^2\right) = \pi a_i^2 \left(-k^2\varphi_{0i}^2 + |\nabla_i\varphi_0|^2\right). \quad (7.119)$$

Now from (7.116)

$$\int_{S_i} (\nabla_i\varphi_0.\boldsymbol{R}_i) \frac{\partial\varphi_0}{\partial R_i}\, \mathrm{d}s \approx \frac{1}{a_i} \int_{S_i} (\nabla_i\varphi_0.\boldsymbol{R}_i)^2\, \mathrm{d}s$$

$$= a_i^2 \,|\nabla_i\varphi_0|^2 \int_0^{2\pi} \cos^2\theta_i\, \mathrm{d}\theta_i = \pi a_i^2 |\nabla_i\varphi_0|^2, \quad (7.120)$$

where θ_i is a polar angle with origin at O_i. The above results when substituted in (7.112) yield

$$k^2 \approx k_0^2 - \frac{\pi \displaystyle\sum_{i=1}^{n} a_i^2 \left(2|\nabla_i\varphi_0|^2 - k_0^2\varphi_{0i}^2\right)}{\displaystyle\iint_{D_0} \varphi_0^2\, \mathrm{d}A}. \quad (7.121)$$

The sloshing frequencies of water within an annular region can be calculated exactly (see McLachlan 1954, §2.64). For an inner pipe of radius b within an outer circle of radius a, the exact solution for the fundamental mode, expressed in terms of the parameter ka, is compared with the approximation (7.121) in Table 7.2. Results corresponding to a uniform distribution of 32 pipes have been calculated by Drake (1999) using the boundary-element method and are compared with the approximation (7.121) in Table 7.3. All pipes have the same radius b and the outer domain is again a circle of radius a. The utility of the present method for this complex geometry is clearly demonstrated.

b/a	ka (exact)	ka (approx)
0.00	1.84118	1.84118
0.05	1.83157	1.83152
0.10	1.80347	1.80220
0.15	1.75977	1.75226
0.20	1.70512	1.67984

Table 7.2 Comparison of exact and approximate values of ka for an annular region.

b/a	ka (numerical)	ka (approx)
0.0000	1.847	1.841
0.0125	1.842	1.836
0.0250	1.825	1.820
0.0375	1.798	1.793
0.0500	1.761	1.754
0.0625	1.716	1.704
0.0750	1.663	1.639

Table 7.3 Comparison of numerical and approximate values of ka for 32 pipes of radius b within a cylinder of radius a. (Reprinted from *Applied Ocean Research*, **3**, K. R. Drake, The effect of internal pipes on the fundamental frequency of liquid sloshing in a circular tank, 133–143, Copyright (1999), with permission from Elsevier Science.)

Bibliographical notes

The informal procedure used in §7.2.1 for the solution of water-wave problems by the method of matched asymptotic expansions was pioneered by Tuck (1975) and applied to the breakwater-gap problem by Mei (1983, §5.5). Although this method works for problems involving an aperture, it fails to give the correct results for certain other problems. For example, the scattering of waves by a rigid semi-infinite duct has an exact solution (Noble 1958, p. 100) which when expanded in an appropriate way does not agree with the result obtained from an informal application of the method of matched asymptotic expansions similar to that in §7.2.1. The reasons for the failure of the matching method in this particular problem are discussed by McIver and Rawlins (1992) and a formal solution is obtained by McIver and Rawlins (1993). The formal matching principle used in the problems discussed here was developed by Crighton and Leppington (1973) and is discussed in detail by Crighton et al. (1992, Chapter 6).

The scattering of long surface waves by various objects has been treated by Davis and Leppington (1977, 1985) with applications in electromagnetic waves and water waves. A survey of long-wave results for the water-wave problem obtained by the method of matched asymptotic expansions is given by McIver (1994a).

Another problem where an informal application of the method of matched asymptotic expansions can give incorrect results is the diffraction of long waves by an obstacle in a wave guide.   This problem is examined in detail by Martin and Dalrymple (1988) who give a procedure for obtaining the correct result. Acoustic wave propagation along a stepped guide with radiation into free space is analyzed by Lesser and Lewis (1972a, 1972b).

As well as the scattering and radiation problems considered here, the method of matched asymptotic expansions may be used to obtain results for eigenvalue problems. For example, McIver (1991) obtains asymptotic results for the trapping of modes by an obstacle in a water-filled channel or, equivalently, an acoustic wave guide.

In addition to the scattering of waves by single structures, the method of matched asymptotic expansions may be used to analyze wave interaction with an array of structures. For example, McIver (1987) considers the scattering of water waves by an array of vertical cylinders and floating hemispheres in order to explain the enhancement of wave forces in long waves. Unlike those discussed here, the problem treated involves three length scales and consequently the fluid domain must be divided into three regions. This work is an application of a technique developed by Balsa (1982, 1983) in the acoustic context to analyze low-frequency oscillations of structures in a flow.

Here we have concentrated on applications of the method of matched asymptotic expansions to problems in which the wavelength is much greater than a typical dimension of the obstacle. The method may also be applied to problems in which the wavelength is much less than a typical obstacle dimension. Such high-frequency applications were pioneered in the field of water waves by Leppington (1972) and, for example, can be used to analyze wave scattering by a partially-immersed horizontal circular cylinder (Alker 1977) and a submerged, horizontal elliptical cylinder (Leppington and Siew 1980). The same method is used by Simon (1985) in an extensive investigation of the high-frequency radiation of waves by surface-piercing structures.

The technique for eigenvalue problems described in §7.5 is based on the work of Curzon and Plant (1986) who analyzed the eigenfrequencies of an acoustic cavity containing a small obstacle. A similar, but more formal, perturbation approach to such problems is described by Mehl and Hill (1989), while the method of matched asymptotic expansions is used by Lesser and Lewis (1974). Similar problems arise in other contexts. For example, Davidovitz and Lo (1987) consider the propagation of electromagnetic waves in an annular guide using similar ideas to Curzon and Plant (1986).

Chapter 8

Variational methods

In §2.5.1 the problem of wave scattering by a vertical barrier was solved using an integral-equation method. Here some of the background to this approach is explained in the context of a variational procedure. In particular, it is explained how the numerical procedure used in §8.1 is able to achieve high accuracy and how complementary bounds may be obtained for physical quantities.

Eigenvalue problems have been used as examples throughout this book. In §8.2 a variational approach to eigenvalue problems in terms of the so-called Rayleigh quotient is described and, in particular, the maximum-minimum definition of eigenvalues is used to obtain some general results. Finally, the Rayleigh-Ritz method for the numerical calculation of eigenvalues is explained in the context of an example from wave propagation through infinite arrays of scatterers.

8.1 Scattering and radiation problems

In a number of scattering and radiation problems it is desired to determine the quantity

$$A = \int_L u(t) f(t) \, \mathrm{d}t, \qquad (8.1)$$

where u solves the integral equation

$$\int_L K(z,t) u(t) \, \mathrm{d}t = f(z), \quad z \in L. \qquad (8.2)$$

Here L is a real interval, the integral operator K is real, positive definite and symmetric, so that $K(z,t) = K(t,z)$, and f is a given real function. One such problem is described in §2.5.1 where the integral equation is (2.99) and the quantity of interest is given in (2.101).

It is convenient to write the integral equation as

$$Ku = f \qquad (8.3)$$

and to introduce the inner product

$$\langle v, w \rangle \equiv \int_L v(t) w(t) \, dt \quad (= \langle w, v \rangle) \qquad (8.4)$$

for real v and w so that

$$A = \langle u, f \rangle. \qquad (8.5)$$

The symmetry of K and the symmetry of the inner product give

$$\langle Kv, w \rangle = \int_L \left\{ \int_L K(z,t) v(t) \, dt \right\} w(z) \, dz$$

$$= \int_L v(t) \left\{ \int_L K(t,z) w(z) \, dz \right\} dt = \langle v, Kw \rangle = \langle Kw, v \rangle \quad (8.6)$$

and the positive definiteness is expressed as

$$\langle Kv, v \rangle \geq 0 \quad \text{for all } v. \qquad (8.7)$$

8.1.1   The variational principle

To obtain a variational principle suitable for the calculation of A, it is first observed that

$$A = \frac{\langle u, f \rangle^2}{\langle u, f \rangle} = \frac{\langle u, f \rangle^2}{\langle Ku, u \rangle}, \qquad (8.8)$$

where the integral equation (8.3) and symmetry of the inner product have been used to rewrite the denominator. Define a functional

$$\mathcal{A}[v] \equiv \frac{\langle v, f \rangle^2}{\langle Kv, v \rangle} \qquad (8.9)$$

and consider the effects of arbitrary small variations in v. To this end write

$$v = v_0 + \epsilon v_1, \quad \epsilon \ll 1, \qquad (8.10)$$

and then

$$A[v_0 + \epsilon v_1] = \frac{\langle v_0, f\rangle^2}{\langle Kv_0, v_0\rangle}\left[1 + 2\epsilon\left(\frac{\langle v_1, f\rangle}{\langle v_0, f\rangle} - \frac{\langle Kv_1, v_0\rangle}{\langle Kv_0, v_0\rangle}\right) + O(\epsilon^2)\right].$$

(8.11)

If v_0 is the solution of (8.3) then

$$\langle Kv_1, v_0\rangle = \langle v_1, Kv_0\rangle = \langle v_1, f\rangle,$$

(8.12)

where the symmetry property (8.6) has been used, and

$$\langle Kv_0, v_0\rangle = \langle f, v_0\rangle = \langle v_0, f\rangle$$

(8.13)

so that the coefficient of ϵ in (8.11) is zero. Conversely, if for all v_1, the coefficient of ϵ is zero then

$$\langle Kv_0, v_0\rangle\langle v_1, f\rangle = \langle v_0, f\rangle\langle Kv_1, v_0\rangle$$

(8.14)

or by (8.6)

$$\langle Kv_0, v_0\rangle\langle f, v_1\rangle = \langle v_0, f\rangle\langle Kv_0, v_1\rangle$$

(8.15)

which holds for any v_1 provided

$$\langle Kv_0, v_0\rangle f = \langle v_0, f\rangle Kv_0.$$

(8.16)

This relation in unaffected by multiplicative scalings of v_0 and so if a scaling is chosen to give $\langle Kv_0, v_0\rangle = \langle v_0, f\rangle$ it follows that $Kv_0 = f$, that is, v_0 satisfies the integral equation (8.3). Thus, it has been demonstrated that $A[v]$ is stationary with respect to small variations in v if and only if v solves (8.3).

8.1.2 Numerical implementation

The variational formulation described above may be used as the basis of a numerical method for the calculation of A. An approximate solution of the integral equation (8.3) is sought in the form

$$U(t) = \sum_{n=0}^{N} a_n u_n(t).$$

(8.17)

The coefficients $\{a_n; n = 0, 1, \ldots, N\}$ are found by exploiting the connection between the solutions of (8.3) and the stationary values of A

and, in particular, by consideration of the changes in $\mathcal{A}[U]$ with respect to variations in the coefficients. A necessary condition for $\mathcal{A}[U]$ to be stationary is that for all $m = 0, 1, \ldots N$

$$\frac{\partial \mathcal{A}}{\partial a_m} = \frac{2\langle U, f \rangle}{\langle KU, U \rangle^2} \left[\langle KU, U \rangle \langle f, u_m \rangle - \langle U, f \rangle \langle KU, u_m \rangle \right] = 0 \qquad (8.18)$$

which is satisfied if

$$\langle KU, U \rangle \langle f, u_m \rangle = \langle U, f \rangle \langle KU, u_m \rangle. \qquad (8.19)$$

This relation is unaffected by multiplicative scalings to U and if a scaling is chosen to give

$$\langle KU, U \rangle = \langle U, f \rangle \qquad (8.20)$$

it reduces to

$$\langle KU, u_m \rangle = \langle f, u_m \rangle, \quad m = 0, 1, \ldots, N, \qquad (8.21)$$

or

$$\sum_{n=0}^{N} a_n \langle K u_n, u_m \rangle = \langle f, u_m \rangle, \quad m = 0, 1, \ldots, N, \qquad (8.22)$$

which is a set of simultaneous equations for the a_n.

Once the coefficients are known then

$$A \approx \langle U, f \rangle = \sum_{n=0}^{N} a_n \langle u_n, f \rangle. \qquad (8.23)$$

From (8.11) it is apparent that the stationarity of $\mathcal{A}[U]$ to small variations in U implies that A can be calculated with an error that is second order in the deviation of U from the true solution u. Further, with the aid of (8.6) and (8.20),

$$\begin{aligned}
\mathcal{A}[u] - \mathcal{A}[U] &= \langle u, f \rangle - \langle U, f \rangle = \langle u, f \rangle - 2\langle U, f \rangle + \langle U, f \rangle \\
&= \langle u, Ku \rangle - 2\langle U, Ku \rangle + \langle KU, U \rangle \\
&= \langle u, Ku \rangle - 2\langle U, Ku \rangle + \langle U, KU \rangle \\
&= \langle U - u, K(U - u) \rangle \\
&\geq 0 \qquad\qquad\qquad\qquad\qquad\qquad\qquad\qquad\qquad (8.24)
\end{aligned}$$

because K is positive definite. Thus

$$\mathcal{A}[U] \leq \mathcal{A}[u] \qquad (8.25)$$

so that $\mathcal{A}[U]$ is a lower bound for $\mathcal{A}[u] \equiv A$.

8.1.3 The Galerkin method

An alternative to the variational solution is a direct solution of the integral equation using the approximation (8.17). Substitution of this approximation into the integral equation gives

$$\sum_{n=0}^{N} a_n K u_n = f \tag{8.26}$$

and then multiplication throughout by each of the $u_m(t)$ and integration over L gives the equations (8.22). This technique, known as the Galerkin method, is therefore entirely equivalent to the variational approach and due to its simplicity is usually preferable. The Galerkin method is used for the solution of the problem in §2.5.1.

8.1.4 Complementary approximations

It is sometimes possible to reformulate the problem defined by (8.3) and (8.5) in terms of a different quantity $p(z)$, $z \in \overline{L}$, where $\overline{L}$ is a 'complementary' interval to the L appearing in §8.1. For example, in the vertical-barrier problem described in §2.5.1, L is the gap beneath the barrier and $\overline{L}$ is the barrier itself. In this reformulation of the problem an integral equation in the form

$$Mp = f \tag{8.27}$$

is obtained, with now

$$A^{-1} = \langle p, f \rangle. \tag{8.28}$$

As for the integral operator K appearing in (8.3), M is real, positive definite, and symmetric so that the theory described above can again be applied.

Define a functional

$$B[p] \equiv \frac{\langle p, f \rangle^2}{\langle Mp, p \rangle}; \tag{8.29}$$

then if

$$P(t) = \sum_{n=0}^{N} b_n p_n(t), \quad t \in \overline{L}, \tag{8.30}$$

is an approximation to p with the b_n determined by the Galerkin method, say, then

$$\mathcal{B}[\mathcal{P}] = \sum_{n=0}^{N} b_n \langle p_n, f \rangle \tag{8.31}$$

is a lower bound for A^{-1}. When combined with the result (8.25) this gives

$$\mathcal{A}[U] \leq A \leq \{\mathcal{B}[P]\}^{-1} \tag{8.32}$$

so that A is bounded above and below and the formulations based on $\mathcal{A}$ and $\mathcal{B}$ are said to yield "complementary approximations" to A.

Bibliographical notes

The theory given here is described in a more general context by Jones (1986, §5.15) and Jones (1994, §4.11 and §5.1) and applications are described in acoustic and electromagnetic waves.

Variational methods of the type described in §8.1.1 have been used for a variety of problem in water waves. In two dimensions, Miles (1967) calculated the transmission past a step on the fluid bottom while Mei and Black (1969) considered the scattering of waves by rectangular obstacles in water of constant depth. Numerical solutions for the scattering of waves by a circular dock, that is, a truncated vertical circular cylinder with fluid beneath it, were obtained by Miles (1971) using variational methods. Evans and Morris (1972) used complementary approximations to show that a pair of identical, surface-piercing, vertical barriers in deep water are able to both totally transmit and totally reflect waves. Complementary approximations were also used by Porter and Evans (1995) in a variety of problems involving thin vertical barriers in finite depth water. Extensive results for both thin and thick vertical barriers are given in Mandal and Chakrabarti (2000).

8.2   Eigenvalue problems

8.2.1   Eigenvalues of the negative Laplacian

Consider a domain D of the x, y-plane surrounded by a boundary S. The aim is to calculate the eigenvalues λ of the problem

$$-\nabla^2 \phi = \lambda \phi \quad \text{in} \quad D, \tag{8.33}$$

$$\frac{\partial \phi}{\partial n} + \sigma \phi = 0 \quad \text{on} \quad S, \tag{8.34}$$

where the real-valued function $\phi(x, y)$ has continuous second derivatives throughout D, n denotes the outward normal to S, and σ is real and piecewise constant on S. For this problem there are an infinity of eigenvalues $\lambda_1 \leq \lambda_2 \leq \ldots \leq \lambda_n \leq \ldots$ all of which are real. The eigenfunctions corresponding to distinct eigenvalues are orthogonal, that is, if ϕ_i and ϕ_j are the eigenfunctions corresponding to the distinct eigenvalues λ_i and λ_j then $\langle \phi_i, \phi_j \rangle = 0$ where here the appropriate inner product is

$$\langle v, w \rangle \equiv \iint_D vw \, dx \, dy. \tag{8.35}$$

Solutions are sought that minimize the "energy functional"

$$\mathcal{E}[v] \equiv \iint_D |\nabla v|^2 \, dx \, dy + \int_S \sigma v^2 \, ds, \tag{8.36}$$

where s is the arc length measured along S, subject to the constraint

$$\mathcal{H}[v] \equiv \iint_D v^2 \, dx \, dy = 1 \tag{8.37}$$

which is used to exclude the trivial solution. This minimization process is equivalent to the minimization of the Rayleigh quotient

$$\mathcal{R}[v] = \frac{\mathcal{E}[v]}{\mathcal{H}[v]}, \tag{8.38}$$

and it will be now be shown how the eigenvalue problem is recovered by examination of the stationary values of $\mathcal{R}$. The test functions v are required to be continuous in the closure of D and to have piecewise first derivatives within D.

Consider a test function of the form

$$v = v_0 + \epsilon v_1, \quad \epsilon \ll 1, \tag{8.39}$$

so that

$$\mathcal{R}[v_0 + \epsilon v_1] - \mathcal{R}[v_0] = \frac{2\epsilon}{\mathcal{H}[v_0]} \left[\iint_D \nabla v_0 \cdot \nabla v_1 \, dx \, dy \right.$$

$$\left. + \sigma \int_S v_0 v_1 \, ds - \mathcal{R}[v_0] \iint_D v_0 v_1 \, dx \, dy \right] + O(\epsilon^2). \tag{8.40}$$

Now by an identity from vector calculus and the divergence theorem

$$\iint_D \left(v_1 \nabla^2 v_0 + \nabla v_1 \cdot \nabla v_0 \right) dx \, dy$$

$$= \iint_D \nabla.(v_1 \nabla v_0) \, dx \, dy = \int_S v_1 \frac{\partial v_0}{\partial n} \, ds \tag{8.41}$$

and so

$$\mathcal{R}[v_0 + \epsilon v_1] - \mathcal{R}[v_0] = \frac{2\epsilon}{\mathcal{H}[v_0]} \left[- \iint_D \left(\nabla^2 v_0 + \mathcal{R}[v_0] v_0 \right) v_1 \, dx \, dy \right.$$

$$\left. + \int_S \left(\frac{\partial v_0}{\partial n} + \sigma v_0 \right) v_1 \, ds \right] + O(\epsilon^2). \tag{8.42}$$

Thus if v_0 is a stationary point of $\mathcal{R}[v_0]$, so that for all v_1 the coefficient of ϵ is zero, then

$$\nabla^2 v_0 + \mathcal{R}[v_0] v_0 = 0 \quad \text{in} \quad D, \tag{8.43}$$

$$\frac{\partial v_0}{\partial n} + \sigma v_0 = 0 \quad \text{on} \quad S, \tag{8.44}$$

and v_0 is an eigenfunction corresponding to the eigenvalue $\lambda = \mathcal{R}[v_0]$.

An important point to note is that the boundary condition (8.34) is a so-called "natural" boundary condition as it is satisfied automatically by functions v_0 corresponding to stationary points of $\mathcal{R}[v]$. In other words,

it is not required to take explicit account of the boundary condition in
the minimization process. However, if a Dirichlet boundary condition
(that is $\phi = 0$) is imposed on S then the final term in (8.36) is dropped
and the test functions must be restricted to those for which $v_0 = v_1 = 0$
on S.

8.2.2   The sloshing problem

Consider a finite three-dimensional fluid domain D with a container
with wetted surface S and bounded above by a free-surface F. The aim
is to calculate the values of λ for which the problem

$$\nabla^2 \phi = 0 \quad \text{in} \quad D, \tag{8.45}$$

$$\frac{\partial \phi}{\partial n} = 0 \quad \text{on} \quad S, \tag{8.46}$$

$$\frac{\partial \phi}{\partial n} = \lambda \phi \quad \text{on} \quad F, \tag{8.47}$$

has non-trivial solutions for the real potential ϕ. In contrast to the
problem in §8.2.1, the eigenvalue now appears in one of the boundary
conditions. For this problem also, there are an infinity of eigenvalues
$\lambda_1 \leq \lambda_2 \leq \ldots \leq \lambda_n \leq \ldots$ all of which are real and the eigenfunctions
corresponding to distinct eigenvalues are orthogonal. That is, if ϕ_i and
ϕ_j are the eigenfunctions corresponding to the distinct eigenvalues λ_i
and λ_j then $\langle \phi_i, \phi_j \rangle = 0$, where now

$$\langle v, w \rangle \equiv \iint_F vw \, \mathrm{d}S. \tag{8.48}$$

The Rayleigh quotient for this problem is

$$\mathcal{R}[v] = \frac{\displaystyle\iiint_V |\nabla v|^2 \, \mathrm{d}V}{\displaystyle\iint_F v^2 \, \mathrm{d}S} \tag{8.49}$$

and a similar calculation to that given in §8.2.1 shows that both of the
boundary conditions (8.46) and (8.47) are natural, and so need not be
satisfied explicitly by a test function v.

8.2.3   The maximum-minimum principle

For both of the problems given in §8.2.1 and §8.2.2 all of the eigenvalues can be characterized in terms of the Rayleigh quotient. The
minimum value of $R[v]$ over all allowable test functions v is the lowest
eigenvalue λ_1 and the minimizing function is the corresponding eigenfunction ϕ_1. For any allowable v, $R[v]$ gives an upper bound for λ_1. The
second eigenvalue λ_2 is obtained by minimizing $R[v]$ with the restriction
that v is orthogonal to ϕ_1, that is $\langle v, \phi_1 \rangle = 0$. The minimum of $R[v]$
when v is orthogonal to ϕ_1 and ϕ_2 is λ_3, and so on.

All eigenvalues can be characterized without reference to the lower
eigenfunctions through the maximum-minimum principle

$$\lambda_n = \max_{W_{n-1}} \left\{ \min_{\substack{\langle v, w_i \rangle = 0 \\ i=1,2,\ldots n-1}} R[v] \right\}, \tag{8.50}$$

where $W_{n-1} = \{w_1, w_2, \ldots, w_{n-1}\}$ is a sequence of admissible test functions. Thus, given the sequence W_{n-1} the Rayleigh quotient $R[v]$ is
minimized over all admissible test functions orthogonal to each member
of W_{n-1}, and then λ_n is obtained by maximizing over all possible se-
quences W_{n-1}. For any sequence W_{n-1} the successive minimum values
$\{\mu_1, \mu_2, \ldots, \mu_n\}$ of $R[v]$ satisfy

$$\mu_1 = \lambda_1, \ \mu_2 \leq \lambda_2, \ \ldots, \ \mu_n \leq \lambda_n \tag{8.51}$$

with equality only if $w_i = \phi_i$, $i = 1, 2, \ldots, n - 1$.

Two important principles follow from the maximum-minimum definition of eigenvalues. These are:

1. Strengthening the conditions in a minimum problem (by imposing
   restrictions on the admissible functions, for example) does not di-
   minish the minimum. Conversely, weakening the conditions does
   not increase the minimum.

   Example (a): The boundary condition (8.34) is natural and test
   functions may be chosen without restriction. If a Dirichlet condition is imposed on some part S_1 of S the test functions must be
   restricted to those that satisfy $v = 0$ on S_1 and hence the eigenvalues will increase (or, at least, not decrease).

   Example (b): For the sloshing problem in §8.2.2, introducing a
   rigid baffle into a container of fluid relaxes the continuity conditions on the admissible functions and the space of test functions

is widened. Hence the eigenvalues will decrease (or at least not increase). The insertion of a rigid vertical baffle into a rectangular container of fluid was studied by Evans and McIver (1987).

2. Given two minimum problems with the same class of admissible functions such that for every admissible function the Rayleigh quotient $\mathcal{R}[u]$ for problem (A) is no smaller than for problem (B), then the minimum for problem (A) is also no smaller than for problem (B).

Example: If the value of σ in the boundary condition (8.34) is increased or decreased then for every admissible v the corresponding $\mathcal{R}[v]$ changes in the same sense. Hence the eigenvalues can change only in the same sense as σ.

8.2.4 The Rayleigh-Ritz method

The relation of eigenvalues to the minima of the Rayleigh quotient is the basis of a numerical method known as the Rayleigh-Ritz method. In the method the trial function v in the Rayleigh quotient $\mathcal{R}[v]$ is taken to be a finite sum of selected functions with unknown coefficients, and the stationary points of $\mathcal{R}[v]$ with respect to variations in these coefficients yield approximations to the eigenvalues. The method is numerically effective because as the eigenvalues are determined from stationary values of $\mathcal{R}[v]$, the error in an eigenvalue is quadratic in the error in the trial function v when compared to the corresponding eigenfunction, as may be seen from the results in §8.2.1.

The method is illustrated here through consideration of the problem of water-wave propagation through an infinite array of rigid vertical cylinders standing in water of constant depth. The cylinders are all of radius c and are arranged in a square array so that every translation between the axes of the cylinders has the form of a so-called lattice vector

$$\boldsymbol{R} = L(m_1\boldsymbol{i} + m_2\boldsymbol{j}), \qquad (8.52)$$

where $\boldsymbol{i}$ and $\boldsymbol{j}$ are unit vector parallel to the x- and y-axes respectively, and m_1 and m_2 are integers. Once the time and depth dependence have been removed, the complex-valued potential ϕ must satisfy the two-dimensional Helmholtz equation

$$(\nabla^2 + k^2)\phi = 0 \qquad (8.53)$$

and the boundary condition

$$\frac{\partial \phi}{\partial n} = 0 \qquad (8.54)$$

on each cylinder, where n is a normal coordinate. Solutions are sought in the form

$$\phi(\boldsymbol{r} + \boldsymbol{R}) = e^{i\boldsymbol{q}\cdot\boldsymbol{r}}\,\phi(\boldsymbol{r}), \qquad (8.55)$$

where $\boldsymbol{R}$ is any lattice vector and $\boldsymbol{r}$ is the position vector of an arbitrary point in the fluid domain. For real $\boldsymbol{q}$ such solutions correspond to waves that propagate through the array without change of amplitude and $\boldsymbol{q}$ measures the changes in the phase of the motion as the array is traversed.

Because of the periodicity of the geometry and the condition (8.55) attention can be restricted to a single square cell of side L containing one centrally-placed cylinder. Denote the fluid domain of this cell by D. If $\boldsymbol{q}$ is specified then the aim is find eigenvalues $\lambda = k^2$ of the negative Laplacian with the boundary condition (8.54) imposed on the cylinder and the periodicity condition (8.55) used to relate the values of the solution on opposite sides of the cell. As for the problems of §8.2.1 and §8.2.2 all of the eigenvalues are real and the eigenfunctions corresponding to distinct eigenvalues are orthogonal where now the inner product

$$\langle v, w \rangle \equiv \iint_D v\overline{w}\,dx\,dy. \qquad (8.56)$$

The energy functional for this problem is

$$\mathcal{E}[v] \equiv \iint_D |\nabla v|^2\,dx\,dy, \qquad (8.57)$$

the constraint is $\mathcal{H}[v] \equiv \langle v, v \rangle = 1$, and as before the Rayleigh quotient $\mathcal{R}[v] = \mathcal{E}[v]/\mathcal{H}[v]$. The condition (8.54) is a natural boundary condition and need not be incorporated into a trial function u; however, the trial functions are required to satisfy (8.55).

The aim is find numerical approximations to the lowest eigenvalues. A trial function for the variational problem is chosen in the form

$$v(x, y) = \sum_{m,n=-N}^{N} a_{mn}\phi_{mn}(\boldsymbol{r}) \qquad (8.58)$$

with coefficients $\{a_{mn};\ m, n = -N, -N+1, \ldots, N\}$ and where the ϕ_{mn} are admissible functions. Here the choice

$$\phi_{mn}(\boldsymbol{r}) = e^{i(\boldsymbol{q}+\boldsymbol{K}_{mn})\cdot\boldsymbol{r}}, \tag{8.59}$$

where

$$\boldsymbol{K}_{mn} = \frac{2\pi}{L}(m\boldsymbol{i} + n\boldsymbol{j}), \tag{8.60}$$

is made in order to enforce (8.55). The minima of the Rayleigh quotient $\mathcal{R}[v]$ as a function of the set $\{a_{mn};\ m, n = -N, -N+1, \ldots, N\}$ yield approximations to the eigenvalues.

As noted in §8.2.3, the minimization of the Rayleigh quotient is equivalent to the minimization of the energy functional $\mathcal{E}[v]$ subject to the constraint $\mathcal{H}[v] = 1$ and this can be achieved by minimizing $\mathcal{E}[v] - \Lambda\mathcal{H}[v]$, where Λ is a Lagrange multiplier. Define

$$E[v, w] \equiv \iint_D \nabla v \cdot \nabla \overline{w} \, dx \, dy \tag{8.61}$$

and

$$H[v, w] \equiv \iint_D v\overline{w} \, dx \, dy \tag{8.62}$$

so that

$$\mathcal{E}[v] - \Lambda\mathcal{H}[v] = \sum_{p,q,m,n=-N}^{N} a_{pq}a_{mn}\left(E[\phi_{pq}, \phi_{mn}] - \Lambda H[\phi_{pq}, \phi_{mn}]\right). \tag{8.63}$$

Necessary conditions for a minimum are obtained by setting the derivative with respect to each of the a_{pq} to zero and this results in the matrix eigenvalue problem

$$(\boldsymbol{E} - \Lambda\boldsymbol{H})\boldsymbol{a} = 0, \tag{8.64}$$

where the $(2N + 1)^2 \times (2N + 1)^2$ matrices have components

$$(\boldsymbol{E})_{pqmn} = E[\phi_{pq}, \phi_{mn}], \quad (\boldsymbol{H})_{pqmn} = H[\phi_{pq}, \phi_{mn}] \tag{8.65}$$

and the $(2N+1)^2$-vector has components $(\boldsymbol{a})_{mn} = a_{mn}$. This eigenvalue problem may be solved by standard numerical methods to determine

N	kL/π					
2	0.452161	1.45616	1.96083	2.12720	2.39254	2.46795
4	0.450747	1.44452	1.95358	2.11008	2.38388	2.45978
6	0.450740	1.44434	1.95334	2.10959	2.38375	2.45942
8	0.450740	1.44434	1.95334	2.10959	2.38374	2.45942

Table 8.1   Eigenvalues kL for wave propagation through an array of circular cylinders illustrating the convergence with increasing values of the truncation parameter N.

approximations to the lowest $P = (2N + 1)^2$ eigenvalues and their corresponding eigenfunctions. If the approximate eigenvalues are arranged in increasing order so that

$$\Lambda_1 \leq \Lambda_2 \leq \cdots \leq \Lambda_P \tag{8.66}$$

then

$$\lambda_n \leq \Lambda_n, \quad n = 1, 2, \ldots P. \tag{8.67}$$

That is, the approximate eigenvalues are upper bounds for the eigenvalues of the original infinite dimensional problem (this follows from the first principle in §8.2.3 as the chosen form of the approximation to the eigenfunction corresponds to a restriction on the space of test functions). The convergence of the method with increasing N is illustrated in Table 8.1 for the lowest six eigenvalues for the case $c/L = 0.5$, $q = (\pi/2)i$. The results are given in terms of the wavenumber $k = \sqrt{\lambda}$.

The example described here is taken from McIver (2000b) where the propagation of such Bloch waves through an array is discussed in some detail.

Bibliographical notes

The development of the theory for the eigenvalues of the Laplacian and the general background and consequences of the maximum-minimum principle are described in detail by Courant and Hilbert (1953, Chapter 6) and Duff and Naylor (1966, Chapter 6). The Rayleigh-Ritz method yields only upper bounds for the eigenvalues, a method that yields both upper and lower bounds is described by Kuttler and Sigillito (1978).

The variational formulation of the sloshing problem is described by Lawrence, Wang, and Reddy (1958) and Moiseev (1964) and a broader mathematical discussion is given by Henrici, Troesch, and Wuytack (1970). Numerous applications of the Rayleigh-Ritz method to sloshing problems are given by Moiseev and Petrov (1965) and a method for computing both upper and lower bounds is described by Fox and Kuttler (1981). A good survey for the sloshing problem of the state of knowledge at the time is given by Fox and Kuttler (1983).

Here we have concentrated on problems where the boundary condition has the form (8.34). An extensive survey of methods for the case of the Dirichlet boundary condition is given by Kuttler and Sigillito (1984).

Appendix A

Bessel functions

This appendix lists some important properties of the Bessel functions J_m, Y_m, I_m, K_m, and $H_m^{(1)}$, where m is an integer, which are used extensively throughout the book. The argument of these functions may be complex and in general we will represent the argument by z. However, if we wish to restrict attention to real arguments only, x will be used as the independent variable. More extensive details of the properties of these functions can be found in many sources, e.g. Watson (1944), McLachlan (1954), Abramowitz and Stegun (1965, Chapter 9).

The Bessel functions $J_m(z)$ and $Y_m(z)$ are linearly-independent solutions of the linear second-order differential equation

$$z^2 \frac{\mathrm{d}^2 f}{\mathrm{d}z^2} + z \frac{\mathrm{d}f}{\mathrm{d}z} + (z^2 - m^2) f = 0 \tag{A.1}$$

and the Hankel function of the first kind, $H_m^{(1)}$, is defined by

$$H_m^{(1)}(z) = J_m(z) + \mathrm{i} Y_m(z). \tag{A.2}$$

The modified Bessel functions $I_m(z)$ and $K_m(z)$ are linearly-independent solutions of the linear second-order differential equation

$$z^2 \frac{\mathrm{d}^2 f}{\mathrm{d}z^2} + z \frac{\mathrm{d}f}{\mathrm{d}z} - (z^2 + m^2) f = 0 \tag{A.3}$$

and they are related to J_m and Y_m by

$$I_m(-\mathrm{i}x) = (-\mathrm{i})^m J_m(x), \tag{A.4}$$

$$K_m(-\mathrm{i}x) = \tfrac{1}{2} \pi \mathrm{i}^{m+1} H_m^{(1)}(x). \tag{A.5}$$

Bessel functions satisfy certain Wronskian relations. For example

$$J_{n+1}(z)Y_n(z) - Y_{n+1}(z)J_n(z) = \frac{2}{\pi z}, \tag{A.6}$$

$$J_n(z)H_n^{(1)\prime}(z) - H_n^{(1)}(z)J_n'(z) = \frac{2i}{\pi z}. \tag{A.7}$$

Asymptotic forms for large real arguments are as follows. As $x \to \infty$,

$$I_m(x) \sim \sqrt{\frac{\pi}{2x}}\, e^x, \tag{A.8}$$

$$K_m(x) \sim \sqrt{\frac{\pi}{2x}}\, e^{-x}, \tag{A.9}$$

$$J_m(x) \sim \sqrt{\frac{2}{\pi x}}\, \cos(x - m\pi/2 - \pi/4), \tag{A.10}$$

$$Y_m(x) \sim \sqrt{\frac{2}{\pi x}}\, \sin(x - m\pi/2 - \pi/4), \tag{A.11}$$

$$H_m^{(1)}(x) \sim \sqrt{\frac{2}{\pi x}}\, e^{i(x - m\pi/2 - \pi/4)}. \tag{A.12}$$

The corresponding results for small real arguments are that as $x \to 0$,

$$I_m(x) \sim (x/2)^m/m!, \tag{A.13}$$

$$K_0(x) \sim -\ln x, \tag{A.14}$$

$$K_m(x) \sim \tfrac{1}{2}(m-1)!\,(x/2)^{-m}, \quad m = 1, 2, 3, \dots, \tag{A.15}$$

$$J_m(x) \sim (x/2)^m/m!, \tag{A.16}$$

$$Y_0(x) \sim -iH_0^{(1)}(x) \sim \frac{2}{\pi}\ln x, \tag{A.17}$$

$$Y_m(x) \sim -iH_m^{(1)}(x)$$
$$\sim \pi^{-1}(m-1)!(x/2)^{-m}, \quad m = 1, 2, 3, \dots. \tag{A.18}$$

For the derivatives of the Bessel functions we have that as $x \to 0$,

$$J_0'(x) \sim -x/2, \tag{A.19}$$

$$J_m'(x) \sim 2^{-m} x^{m-1}/(m-1)!, \quad m = 1, 2, 3, \dots, \tag{A.20}$$

$$H_0^{(1)\prime}(x) \sim 2i/\pi x, \tag{A.21}$$

$$H_m^{(1)\prime}(x) \sim 2^m i m!/\pi x^{m+1}, \quad m = 1, 2, 3, \dots. \tag{A.22}$$

The functions $J_m(z)$, $Y_m(z)$, $J_m'(z)$, and $Y_m'(z)$, $m = 0, 1, 2, \dots$, all possess an infinite number of real zeros. For J_m, Y_m, Y_m', $m \geq 0$, and for J'_m, $m > 0$, the nth positive zeros are labelled j_{mn}, y_{mn}, j'_{mn}, and y'_{mn}, respectively, but $z = 0$ is counted as the first zero of $J'_0(z)$. The zeros interlace:

$$j_{m1} < j_{m+1,1} < j_{m2} < j_{m+1,2} < j_{m3} < \cdots \qquad (A.23)$$

$$y_{m1} < y_{m+1,1} < y_{m2} < j_{m+1,2} < y_{m3} < \cdots \qquad (A.24)$$

$$m \leq j'_{m1} < y_{m1} < y'_{m1} < j_{m1} < j'_{m2} < y_{m2} < y'_{m2} < j_{m2} < j'_{m3} < \cdots . \qquad (A.25)$$

For fixed m and as $n \to \infty$,

$$j_{mn} \sim s - \frac{\mu - 1}{8s} - \frac{4(\mu - 1)(7\mu - 31)}{3(8s)^3}, \qquad (A.26)$$

$$j'_{mn} \sim t - \frac{\mu + 3}{8t} - \frac{4(7\mu^2 + 82\mu - 9)}{3(8t)^3}, \qquad (A.27)$$

where $\mu = 4m^2$, $s = (n + m/2 - 1/4)\pi$ and $t = (n + m/2 - 3/4)\pi$. Exactly the same asymptotic expansions apply to the zeros y_{mn} and y'_{mn} if the values of s and t are interchanged.

Appendix B

Multipoles

This appendix lists formulas for, and properties of, the multipole potentials used in Chapter 3 and various source potentials used throughout the book. Derivations of the formulas which follow are not provided; for these the reader is directed to, for example, Thorne (1953), Wehausen and Laitone (1960), and Mei (1983). Some of the expressions which follow cannot be found in any of these standard sources and, if they are not simply obvious extensions, an explicit reference is provided.

B.1 Two dimensions, infinite depth

In this section we consider solutions of Laplace's equation in two dimensions which are singular at $x = \xi$, $z = \zeta \leq 0$ and which satisfy the free surface boundary condition (1.13). The coordinates that are used are illustrated in Figure B.1 and defined below:

$$r = [(x - \xi)^2 + (z - \zeta)^2]^{1/2}, \qquad r_1 = [(x - \xi)^2 + (z + \zeta)^2]^{1/2},$$
$$X = x - \xi = r \sin \theta = r_1 \sin \theta_1,$$
$$\zeta - z = r \cos \theta, \qquad \zeta + z = r_1 \cos \theta_1.$$

Solutions to $\nabla^2 \phi = 0$ which are singular at $r = 0$ are $\ln r$, $r^{-n} \cos n\theta$ and $r^{-n} \sin n\theta$, $n = 1, 2, 3, \ldots$. If the singularity at (ξ, ζ) is of the form $\ln r$ then the potential corresponds to a pulsating source of fluid and such functions are considered first, with a distinction being made between functions whose singularity lies in the undisturbed free surface $z = 0$ and those whose singularity is submerged. Multipoles with higher order

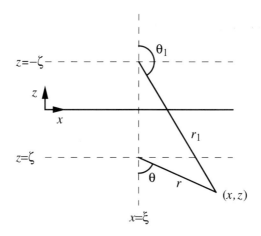

FIGURE B.1
Coordinates used in two dimensions for the infinite depth case.

submerged singularities are treated next and are divided into those which are symmetric about $x = \xi$ and those which are antisymmetric about this line. Multipoles with singularities in the free surface (other than sources) are not presented, but the method for constructing them is essentially the same as that for the source case. All these potentials behave like outgoing waves as $|x| \to \infty$, but we can also construct so-called wave-free potentials which decay at infinity and these are considered last.

In order to aid the reader in manipulating the expressions that are given we note the following integral representations of the singularities described above (see Gradshteyn and Ryzhik 1980, eqns 3.943, 3.944(5) and (6)):

$$\ln r = \int_0^\infty (e^{-\mu} - e^{-\mu|z-\zeta|} \cos \mu X) \frac{d\mu}{\mu}, \tag{B.1}$$

$$\frac{\cos n\theta}{r^n} = \frac{1}{(n-1)!} \int_0^\infty \mu^{n-1} e^{\mu(z-\zeta)} \cos \mu X \, d\mu, \qquad z < \zeta, \tag{B.2}$$

$$= \frac{(-1)^n}{(n-1)!} \int_0^\infty \mu^{n-1} e^{\mu(\zeta-z)} \cos \mu X \, d\mu, \qquad z > \zeta, \tag{B.3}$$

the representations of $r^{-n} \sin n\theta$ being given by (B.2) and (B.3) with cos replaced by sin and an extra factor of -1 in the expression valid for $z > \zeta$.

Submerged source

A submerged source situated at (ξ, ζ), $\zeta < 0$ can be written

$$\phi = \ln \frac{r}{r_1} - 2\!\!\!\!\!\!\diagup\!\!\!\!\!\!\int_0^\infty \frac{e^{\mu(z+\zeta)}}{\mu - K} \cos \mu X \, d\mu \tag{B.4}$$

$$= \ln r - \diagup\!\!\!\!\!\!\int_0^\infty \left\{ \frac{\mu + K}{\mu - K} e^{\mu(z+\zeta)} \cos \mu X + e^{-\mu} \right\} \frac{d\mu}{\mu} \tag{B.5}$$

$$\sim -2\pi i \, e^{K(z+\zeta)} e^{\pm iKX} \quad \text{as} \quad x \to \pm\infty. \tag{B.6}$$

The second of these formulas is given incorrectly by Wehausen and Laitone (1960, p. 482). The imaginary part of ϕ is given explicitly by

$$\operatorname{Im} \phi = -2\pi \, e^{K(z+\zeta)} \cos KX. \tag{B.7}$$

Alternatively, provided $x \neq \xi$, we may write

$$\phi + 2\pi i \, e^{K(z+\zeta)} e^{iK|X|}$$

$$= \ln \frac{r}{r_1} - 2 \int_0^\infty \frac{\mu \cos \mu(z+\zeta) + K \sin \mu(z+\zeta)}{\mu^2 + K^2} e^{-\mu|X|} \, d\mu \tag{B.8}$$

$$= -2 \int_0^\infty \frac{(K \sin \mu z + \mu \cos \mu z)(K \sin \mu\zeta + \mu \cos \mu\zeta)}{\mu(\mu^2 + K^2)} e^{-\mu|X|} \, d\mu \tag{B.9}$$

and a power series expansion can be obtained from (B.4):

$$\phi = \ln \frac{r}{r_1} + \sum_{m=0}^\infty A_m r^m \cos m\theta, \qquad r < 2|\zeta|, \tag{B.10}$$

where

$$A_m = -\frac{2(-1)^m}{m!} \diagup\!\!\!\!\!\!\int_0^\infty \frac{\mu^m e^{2\mu\zeta}}{\mu - K} \, d\mu. \tag{B.11}$$

Another expression valid for $x \neq \xi$, due to Kim (1965) and useful for numerical calculations, is

$$\phi = \ln r + \ln r_1 + 2 e^{K(z+\zeta)} \left[-\ln |X| \right.$$

$$+ (\operatorname{Ci}(K|X|) - i\pi) \cos KX + \left(\operatorname{Si}(K|X|) + \frac{\pi}{2} \right) \sin K|X|$$

$$\left. - \int_{z+\zeta}^0 e^{-K\mu} \ln(X^2 + \mu^2)^{1/2} \, d\mu \right], \tag{B.12}$$

where Ci and Si are the sine and cosine integrals, respectively (Abramowitz and Stegun 1965, §5.2).

Free-surface source

Expressions for a source situated in the free surface at $x = \xi$ can be obtained from (B.4) and (B.8) by setting $r = r_1$, $\zeta = 0$ and (so that we still have $\phi \sim \ln r$ as $r \to 0$) dividing the result by two. Thus

$$\phi = -\fint_0^\infty \frac{e^{\mu z}}{\mu - K} \cos \mu X \, d\mu \tag{B.13}$$

$$= -\pi i\, e^{Kz}\, e^{iK|X|} - \int_0^\infty \frac{\mu \cos \mu z + K \sin \mu z}{\mu^2 + K^2}\, e^{-\mu|X|} \, d\mu \qquad (x \neq \xi) \tag{B.14}$$

$$\sim -\pi i\, e^{Kz}\, e^{\pm iKX} \quad \text{as} \quad x \to \pm\infty. \tag{B.15}$$

The series (B.10) is not valid when $\zeta = 0$, but a power series expansion can be obtained from (B.13) as shown in Yu and Ursell (1961). Throughout the fluid, except when $r = 0$,

$$\phi = e^{Kz}\left[(\gamma + \ln Kr - \pi i) \cos Kx + \theta \sin Kx\right]$$
$$- \sum_{m=1}^\infty \frac{(-Kr)^m}{m!} \left(1 + \tfrac{1}{2} + \ldots + \tfrac{1}{m}\right) \cos m\theta, \tag{B.16}$$

where $\gamma = 0.5772\ldots$ is Euler's constant. The right-hand side of (B.16) can be completely expanded in powers of r by noting that $\exp(Kz + iKX) = \exp(-Kr\, e^{-i\theta})$.

Symmetric multipoles

Multipoles which are singular at (ξ, ζ), $\zeta < 0$, and symmetric about $x = \xi$ can be written

$$\phi_n = \frac{\cos n\theta}{r^n} + \frac{(-1)^n}{(n-1)!} \fint_0^\infty \frac{\mu + K}{\mu - K} \mu^{n-1}\, e^{\mu(z+\zeta)} \cos \mu X \, d\mu \tag{B.17}$$

$$\sim \frac{2\pi i(-K)^n}{(n-1)!}\, e^{K(z+\zeta)}\, e^{\pm iKX} \quad \text{as} \quad x \to \pm\infty. \tag{B.18}$$

The imaginary part of ϕ_n is given explicitly by

$$\operatorname{Im}\phi_n = \frac{2\pi(-K)^n}{(n-1)!}\, e^{K(z+\zeta)} \cos KX. \tag{B.19}$$

Alternatively, provided $x \neq \xi$, we may write

$$
\phi_n - \frac{2\pi i(-K)^n}{(n-1)!} e^{K(z+\zeta)} e^{iK|X|}
$$

$$
= \frac{\cos n\theta}{r^n} + \frac{(-1)^n}{(n-1)!} \int_0^\infty \frac{\mu^{n-1} e^{-\mu|X|}}{\mu^2 + K^2} g_n(\mu)\, d\mu \tag{B.20}
$$

$$
= \frac{2}{(n-1)!} \int_0^\infty \frac{\mu^{n-1} e^{-\mu|X|}}{\mu^2 + K^2} (\mu \cos \mu z + K \sin \mu z) f_n(\mu)\, d\mu, \tag{B.21}
$$

where

$$
g_n(\mu) = \operatorname{Re}\left[i^n(\mu^2 - 2i\mu K - K^2) e^{i\mu(z+\zeta)} \right] \tag{B.22}
$$

and

$$
f_n(\mu) = \begin{cases} (-1)^{n/2}(\mu \cos \mu\zeta + K \sin \mu\zeta) & n \text{ even,} \\ (-1)^{(n-1)/2}(\mu \sin \mu\zeta - K \cos \mu\zeta) & n \text{ odd.} \end{cases} \tag{B.23}
$$

A power series expansion can be obtained from (B.17):

$$
\phi_n = \frac{\cos n\theta}{r^n} + \sum_{m=0}^\infty A_{mn} r^m \cos m\theta, \qquad r < 2|\zeta|, \tag{B.24}
$$

where

$$
A_{mn} = \frac{(-1)^{m+n}}{m!(n-1)!} \int_0^\infty \frac{\mu + K}{\mu - K} \mu^{m+n-1} e^{2\mu\zeta}\, d\mu. \tag{B.25}
$$

Antisymmetric multipoles

Multipoles which are singular at (ξ, ζ), $\zeta < 0$, and antisymmetric about $x = \xi$ can be written

$$
\phi_n = \frac{\sin n\theta}{r^n} - \frac{(-1)^n}{(n-1)!} \int_0^\infty \frac{\mu + K}{\mu - K} \mu^{n-1} e^{\mu(z+\zeta)} \sin \mu X\, d\mu \tag{B.26}
$$

$$
\sim \mp \frac{2\pi(-K)^n}{(n-1)!} e^{K(z+\zeta)} e^{\pm iKX} \qquad \text{as} \quad x \to \pm\infty. \tag{B.27}
$$

The imaginary part of ϕ_n is given explicitly by

$$
\operatorname{Im} \phi_n = -\frac{2\pi(-K)^n}{(n-1)!} e^{K(z+\zeta)} \sin KX. \tag{B.28}
$$

Alternatively, provided $x \neq \xi$, we can write

$$\phi_n + \frac{2\pi(-K)^n}{(n-1)!} \operatorname{sgn}(X) e^{K(z+\zeta)} e^{iK|X|}$$

$$= \frac{\sin n\theta}{r^n} - \frac{(-1)^n \operatorname{sgn}(X)}{(n-1)!} \int_0^\infty \frac{\mu^{n-1} e^{-\mu|X|}}{\mu^2 + K^2} g_n(\mu) \, d\mu \qquad (B.29)$$

$$= \frac{2 \operatorname{sgn}(X)}{(n-1)!} \int_0^\infty \frac{\mu^{n-1} e^{-\mu|X|}}{\mu^2 + K^2} (\mu \cos \mu z + K \sin \mu z) f_n(\mu) \, d\mu, \quad (B.30)$$

where

$$g_n(\mu) = \operatorname{Im} \left[i^n (\mu^2 - 2i\mu K - K^2) e^{i\mu(z+\zeta)} \right] \qquad (B.31)$$

and

$$f_n(\mu) = \begin{cases} (-1)^{n/2}(-\mu \sin \mu\zeta + K \cos \mu\zeta) & n \text{ even}, \\ (-1)^{(n-1)/2}(\mu \cos \mu\zeta + K \sin \mu\zeta) & n \text{ odd}. \end{cases} \qquad (B.32)$$

A power series expansion can be obtained from (B.26):

$$\phi_n = \frac{\sin n\theta}{r^n} + \sum_{m=1}^\infty A_{mn} r^m \sin m\theta, \qquad r < 2|\zeta|, \qquad (B.33)$$

where A_{mn} is again given by (B.25).

Wave-free potentials

It is clear from (B.18) and (B.27) that in both the symmetric and anti-symmetric cases, the combination $\psi_n = \phi_{n+1} + Kn^{-1}\phi_n$, $n = 1, 2, 3, \ldots$ corresponds to a wave-free singularity. In the symmetric case we obtain

$$\psi_n^s = \frac{\cos(n+1)\theta}{r^{n+1}} + \frac{K}{n} \frac{\cos n\theta}{r^n} + \frac{\cos(n+1)\theta_1}{r_1^{n+1}} - \frac{K}{n} \frac{\cos n\theta_1}{r_1^n} \qquad (B.34)$$

and in the antisymmetric case

$$\psi_n^a = \frac{\sin(n+1)\theta}{r^{n+1}} + \frac{K}{n} \frac{\sin n\theta}{r^n} - \frac{\sin(n+1)\theta_1}{r_1^{n+1}} + \frac{K}{n} \frac{\sin n\theta_1}{r_1^n}. \qquad (B.35)$$

When the singularity is in the free surface these reduce to $\psi_{2n}^{\mathrm{s}} \equiv 0$, $\psi_{2n-1}^{\mathrm{a}} \equiv 0$ and (see Ursell 1949)

$$\psi_{2n-1}^{\mathrm{s}} = \frac{\cos 2n\theta}{r^{2n}} + \frac{K}{2n-1}\frac{\cos(2n-1)\theta}{r^{2n-1}}, \qquad (\text{B.36})$$

$$\psi_{2n}^{\mathrm{a}} = \frac{\sin(2n+1)\theta}{r^{2n+1}} - \frac{K}{2n}\frac{\sin 2n\theta}{r^{2n}}. \qquad (\text{B.37})$$

B.2 Two dimensions, finite depth

In this section we consider solutions of Laplace's equation in two dimensions which are singular at $x = \xi$, $z = \zeta \le 0$ and which satisfy both the free-surface boundary condition (1.13) and the bed boundary condition (2.1). The coordinates that are used are illustrated in Figure B.2 and we use the same notation as in §B.1 together with

$$r_2 = [(x-\xi)^2 + (z+\zeta+2h)^2]^{1/2}, \quad z_> = \max(z,\zeta), \quad z_< = \min(z,\zeta).$$

The definitions of k and k_m, N_m, $m = 0, 1, 2, \ldots$ are those of §2.1.

Submerged source

A submerged source situated at $x = (\xi, \zeta)$, $\zeta < 0$ can be written

$$\phi = \ln\frac{r}{r_1} - 2\int_0^\infty \frac{\cos\mu X}{\cosh\mu h}$$

$$\times \left[\frac{\cosh\mu(z+h)\cosh\mu(\zeta+h)}{\mu\sinh\mu h - K\cosh\mu h} + \frac{e^{-\mu h}}{\mu}\sinh\mu z\sinh\mu\zeta\right]\mathrm{d}\mu \quad (\text{B.38})$$

$$\sim -\frac{\pi i}{khN_0^2}\cosh k(z+h)\cosh k(\zeta+h)\,e^{\pm ikX} \quad \text{as} \quad x \to \pm\infty. \quad (\text{B.39})$$

The imaginary part of ϕ is given explicitly by

$$\operatorname{Im}\phi = -\frac{\pi}{khN_0^2}\cosh k(z+h)\cosh k(\zeta+h)\cos kX. \qquad (\text{B.40})$$

Two alternative integral representations are

$$\phi = \ln\frac{r}{h} + \ln\frac{r_2}{h} - 2\int_0^\infty \frac{e^{-\mu h}}{\mu}$$

$$\times \left[\frac{(\mu+K)\cosh\mu(z+h)\cosh\mu(\zeta+h)}{\mu\sinh\mu h - K\cosh\mu h}\cos\mu X + 1\right]\mathrm{d}\mu \qquad (\text{B.41})$$

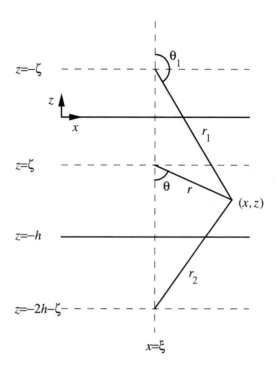

FIGURE B.2
Coordinates used in two dimensions for the finite depth case.

(this formula is given incorrectly in both Wehausen and Laitone 1960, eqn 13.34, and Mei 1983, p. 382) and

$$\phi = -2\int_0^\infty \frac{K\sinh\mu z_> + \mu\cosh\mu z_>}{\mu\sinh\mu h - K\cosh\mu h}\cosh\mu(z_< + h)\frac{\cos\mu X}{\mu}\,\mathrm{d}\mu. \quad \text{(B.42)}$$

Provided $x \neq \xi$, ϕ can be expanded as a series of eigenfunctions:

$$\phi = -\sum_{m=0}^\infty \frac{\pi}{k_m h N_m^2}\cos k_m(z+h)\cos k_m(\zeta + h)\,\mathrm{e}^{-k_m|X|}. \quad \text{(B.43)}$$

Free-surface source

An expression for a source situated in the free surface at $x = \xi$ can be obtained from (B.38) by setting $r = r_1$, $\zeta = 0$ and (so that we still have $\phi \sim \ln r$ as $r \to 0$) dividing the result by two. Alternatively we may write

$$
\phi = -\int_0^\infty \frac{e^{\mu z}}{\mu - K} \cos \mu X \, d\mu
$$
$$
+ \int_0^\infty \frac{e^{-\mu h}(K \sinh \mu z + \mu \cosh \mu z)}{(\mu - K)(K \cosh \mu h - \mu \sinh \mu h)} \cos \mu X \, d\mu, \quad \text{(B.44)}
$$

where the first term is just the infinite depth free-surface source (B.13). The imaginary part of ϕ is given explicitly by

$$
\mathrm{Im}\,\phi = -\frac{\pi i}{2kh N_0^2} \cosh kh \cosh k(z + h) \cos kX \quad \text{(B.45)}
$$

and

$$
\phi \sim -\frac{\pi i}{2kh N_0^2} \cosh kh \cosh k(z + h) \, e^{\pm ikX} \quad \text{as} \quad x \to \pm\infty. \quad \text{(B.46)}
$$

A power series expansion valid for $|z| < 2h$ ($r \neq 0$) can be obtained from (B.44) as shown in Yu and Ursell (1961). The expansion is exactly as for the infinite depth case, (B.16), with the addition of the term

$$
\sum_{m=0}^\infty A_m r^m \cos m\theta,
$$

where

$$
A_{2s} = -\frac{1}{(2s)!} \fint_0^\infty \frac{\mu^{2s+1} \, e^{-\mu h}}{(\mu - K)(\mu \sinh \mu h - K \cosh \mu h)} d\mu, \quad \text{(B.47)}
$$

$$
A_{2s+1} = \frac{K}{(2s+1)!} \fint_0^\infty \frac{\mu^{2s+1} \, e^{-\mu h}}{(\mu - K)(\mu \sinh \mu h - K \cosh \mu h)} d\mu. \quad \text{(B.48)}
$$

Symmetric multipoles

Multipoles which are singular at (ξ, ζ), $\zeta < 0$, and symmetric about $x = \xi$ can be written

$$
\phi_n = \frac{\cos n\theta}{r^n} + \frac{1}{(n-1)!} \fint_0^\infty \frac{\mu^{n-1} \cos \mu X}{\mu \sinh \mu h - K \cosh \mu h} f_n(\mu) \, d\mu, \quad \text{(B.49)}
$$

where

$$
f_n(\mu) = e^{-\mu(\zeta + h)}(K \sinh \mu z + \mu \cosh \mu z)
$$
$$
+ (-1)^n (\mu + K) \, e^{\mu \zeta} \cosh \mu(z + h). \quad \text{(B.50)}
$$

An expression for ϕ_n as a single integral, analogous to (B.42), can be obtained by using the integral representations (B.2) and (B.3); the form of the resulting expression depends on whether n is odd or even. This could then be used to derive an eigenfunction expansion for ϕ_n, analogous to (B.43), if required. Similar remarks pertain to the antisymmetric multipoles discussed below.

The imaginary part of ϕ_n is given explicitly by

$$\operatorname{Im}\phi_n = \frac{\pi k^{n-1}\cosh k(z+h)}{2hN_0^2(n-1)!}\left(e^{-k(\zeta+h)} +(-1)^n e^{k(\zeta+h)}\right)\cos kX \tag{B.51}$$

and as $x \to \pm\infty$

$$\phi_n \sim \frac{\pi i k^{n-1}\cosh k(z+h)}{2hN_0^2(n-1)!}\left(e^{-k(\zeta+h)} +(-1)^n e^{k(\zeta+h)}\right)e^{\pm ikX} . \tag{B.52}$$

A power series expansion can be obtained from (B.49):

$$\phi_n = \frac{\cos n\theta}{r^n} + \sum_{m=0}^{\infty} A_{mn}r^m \cos m\theta, \qquad r < 2|\zeta|, \tag{B.53}$$

where

$$A_{mn} = \frac{1}{2m!(n-1)!}\int_0^{\infty} \frac{\mu^{m+n-1}}{\mu\sinh\mu h - K\cosh\mu h}g_{mn}(\mu)\,\mathrm{d}\mu \tag{B.54}$$

with

$$g_{mn}(\mu) = (\mu + K)\left((-1)^{m+n} e^{\mu(2\zeta+h)} +((-1)^m + (-1)^n)e^{-\mu h}\right)$$
$$+ (\mu - K)e^{-\mu(2\zeta+h)} . \tag{B.55}$$

Antisymmetric multipoles

Multipoles which are singular at (ξ,ζ), $\zeta < 0$, and antisymmetric about $x = \xi$ can be written

$$\phi_n = \frac{\sin n\theta}{r^n} + \frac{1}{(n-1)!}\int_0^{\infty} \frac{\mu^{n-1}\sin\mu X}{\mu\sinh\mu h - K\cosh\mu h}f_n(\mu)\,\mathrm{d}\mu, \tag{B.56}$$

where

$$f_n(\mu) = e^{-\mu(\zeta+h)}(K\sinh\mu z + \mu\cosh\mu z)$$
$$- (-1)^n(\mu + K)e^{\mu\zeta}\cosh\mu(z + h). \tag{B.57}$$

The imaginary part of ϕ_n is given explicitly by

$$\operatorname{Im}\phi_n = \frac{\pi k^{n-1}\cosh k(z+h)}{2hN_0^2(n-1)!}\left(e^{-k(\zeta+h)} -(-1)^n\,e^{k(\zeta+h)}\right)\sin kX \tag{B.58}$$

and as $x\to\pm\infty$

$$\phi_n \sim \pm\frac{\pi k^{n-1}\cosh k(z+h)}{2hN_0^2(n-1)!}\left(e^{-k(\zeta+h)} -(-1)^n\,e^{k(\zeta+h)}\right)e^{\pm ikX}. \tag{B.59}$$

A power series expansion can be obtained from (B.56):

$$\phi_n = \frac{\sin n\theta}{r^n} + \sum_{m=1}^{\infty} A_{mn}r^m\sin m\theta, \qquad r < 2|\zeta|, \tag{B.60}$$

where

$$A_{mn} = \frac{1}{2m!(n-1)!}\int_0^\infty \frac{\mu^{m+n-1}}{\mu\sinh\mu h - K\cosh\mu h}g_{mn}(\mu)\,d\mu \tag{B.61}$$

with

$$g_{mn}(\mu) = (\mu + K)\left((-1)^{m+n}\,e^{\mu(2\zeta+h)} -((-1)^m + (-1)^n)\,e^{-\mu h}\right)$$
$$+ (\mu - K)\,e^{-\mu(2\zeta+h)}. \tag{B.62}$$

Wave-free potentials

Wave-free singularities can be constructed as for the infinite depth case. Equation (B.52) shows that in the symmetric case the potentials

$$\phi_{2n} + \frac{k}{2n-1}\coth k(\zeta+h)\phi_{2n-1}, \qquad \phi_{2n+1} + \frac{k}{2n}\tanh k(\zeta+h)\phi_{2n}$$

are wave free and it follows from (B.59) that in the antisymmetric case tanh and coth should be interchanged in these formulas.

B.3   Three dimensions, infinite depth

In this section we consider solutions of Laplace's equation in three dimensions which are singular at $x = \xi$, $y = \eta$, $z = \zeta \leq 0$ and which satisfy the free-surface boundary condition (1.13). We use both spherical coordinates (r, θ, α), (r_1, θ_1, α) and cylindrical coordinates (R, α, z) defined by

$$r = [(x - \xi)^2 + (y - \eta)^2 + (z - \zeta)^2]^{1/2},$$
$$r_1 = [(x - \xi)^2 + (y - \eta^2) + (z + \zeta)^2]^{1/2},$$
$$\zeta - z = r \cos\theta, \quad \zeta + z = r_1 \cos\theta_1,$$
$$R = [(x - \xi)^2 + (y - \eta^2)]^{1/2} = r \sin\theta = r_1 \sin\theta_1,$$
$$x - \xi = R \cos\alpha, \quad y - \eta = R \sin\alpha.$$

Many of the formulas below involve the associate Legendre functions $P_n^m(x)$. The definition of these functions that we use in this book is

$$P_n^m(x) = (-1)^m (1 - x^2)^{m/2} \frac{\mathrm{d}^m P_n(x)}{\mathrm{d}x^m},$$

which corresponds to that used in Abramowitz and Stegun (1965) and Gradshteyn and Ryzhik (1980), but differs by a factor of $(-1)^m$ from that used by Thorne (1953) and some other authors.

Solutions to Laplace's equation in three dimensions which are singular at $r = 0$ are $r^{-n-1} P_n^m(\cos\theta) \cos m\alpha$, $n \geq m \geq 0$. We note the following integral representations of these singularities:

$$\frac{P_n^m(\cos\theta)}{r^{n+1}} = \frac{(-1)^m}{(n-m)!} \int_0^\infty \mu^n e^{\mu(z-\zeta)} J_m(\mu R) \, \mathrm{d}\mu, \quad z < \zeta, \qquad \text{(B.63)}$$

$$= \frac{(-1)^n}{(n-m)!} \int_0^\infty \mu^n e^{\mu(\zeta-z)} J_m(\mu R) \, \mathrm{d}\mu, \quad z > \zeta. \qquad \text{(B.64)}$$

The special case $n = m = 0$ corresponds to a $1/r$ singularity which represents a pulsating source of fluid and many of the expressions for such a source can be obtained by simply setting $n = m = 0$ in the corresponding expressions for the more general multipoles. The source is such an important function, however, that it is worth describing it separately.

Submerged source

A submerged source situated at (ξ, η, ζ), $\zeta < 0$, can be written

$$\phi = \frac{1}{r} + \int_0^\infty \frac{\mu + K}{\mu - K}\, e^{\mu(z+\zeta)}\, J_0(\mu R)\mathrm{d}\mu \tag{B.65}$$

$$= \frac{1}{r} + \frac{1}{r_1} + \int_0^\infty \frac{2K}{\mu - K}\, e^{\mu(z+\zeta)}\, J_0(\mu R)\mathrm{d}\mu \tag{B.66}$$

$$\sim 2\pi i K\, e^{K(z+\zeta)}\, H_0^{(1)}(KR) \quad \text{as} \quad R \to \infty. \tag{B.67}$$

The imaginary part of ϕ is given explicitly by (B.78) with $n = m = 0$. Alternatively, if $R > 0$,

$$\phi - 2\pi i K\, e^{K(z+\zeta)}\, H_0^{(1)}(KR)$$
$$= \frac{1}{r} + \frac{2}{\pi} \int_0^\infty \frac{(\mu^2 - K^2)\cos\mu(z+\zeta) + 2\mu K \sin\mu(z+\zeta)}{\mu^2 + K^2} K_0(\mu R)\, \mathrm{d}\mu \tag{B.68}$$

$$= \frac{1}{r} + \frac{1}{r_1} - \frac{4K}{\pi} \int_0^\infty \frac{K \cos\mu(z+\zeta) - \mu \sin\mu(z+\zeta)}{\mu^2 + K^2} K_0(\mu R)\, \mathrm{d}\mu. \tag{B.69}$$

A power series expansion is given by (B.81), (B.82) with $n = m = 0$.

Another expression valid for $R > 0$, due to Kim (1965) but implicit in the work of Havelock (1955) and useful for numerical calculations, is

$$\phi = \frac{1}{r} + \frac{1}{r_1} - K\, e^{K(z+\zeta)} \left[\pi\left(\mathbf{H}_0(KR) + Y_0(KR) - 2\mathrm{i}J_0(KR)\right) \right.$$
$$\left. + 2 \int_{z+\zeta}^0 \frac{e^{-K\mu}}{(R^2 + \mu^2)^{1/2}}\, \mathrm{d}\mu \right], \quad \text{(B.70)}$$

where $\mathbf{H}_0$ is the Struve function (Abramowitz and Stegun 1965, Chapter 12).

Free-surface source

A source situated at $(\xi, \eta, 0)$ can be written

$$\phi = \frac{1}{r} + \int_0^\infty \frac{K\, e^{\mu z}}{\mu - K} J_0(\mu R)\mathrm{d}\mu = \int_0^\infty \frac{\mu\, e^{\mu z}}{\mu - K} J_0(\mu R)\mathrm{d}\mu \tag{B.71}$$

$$\sim \pi \mathrm{i} K\, e^{Kz}\, H_0^{(1)}(KR) \quad \text{as} \quad R \to \infty. \tag{B.72}$$

The imaginary part of ϕ is given explicitly by

$$\text{Im}\,\phi = \pi K\, e^{K(z+\zeta)}\, J_0(KR). \tag{B.73}$$

An expansion in terms of spherical harmonics, valid for all $r > 0$, can be obtained as shown in Hulme (1982):

$$\phi = \frac{1}{r} + \pi i K \sum_{n=0}^{\infty} \frac{(-Kr)^n}{n!} P_n(\cos\theta)$$

$$- K \sum_{n=0}^{\infty} (-1)^n \frac{\partial}{\partial \nu} \left\{ \frac{(Kr)^\nu}{\nu!} P_\nu(\cos\theta) \right\}_{\nu=n}. \tag{B.74}$$

Multipoles

Multipoles singular at (ξ, η, ζ), $\zeta < 0$, are of the form $\phi_n^m \cos m\alpha$ and $\phi_n^m \sin m\alpha$, where

$$\phi_n^m = \frac{P_n^m(\cos\theta)}{r^{n+1}} + \frac{(-1)^n}{(n-m)!} \oint_0^\infty \frac{\mu+K}{\mu-K} \mu^n\, e^{\mu(z+\zeta)}\, J_m(\mu R)\mathrm{d}\mu \tag{B.75}$$

$$\sim \frac{2\pi i(-1)^n}{(n-m)!} K^{n+1}\, e^{K(z+\zeta)}\, H_m^{(1)}(KR) \quad \text{as} \quad R \to \infty \tag{B.76}$$

$$\sim \frac{2\pi i(-1)^n}{(n-m)!} K^{n+1}\, e^{K(z+\zeta)} \left(\frac{2}{\pi KR}\right)^{1/2} e^{i(KR - \frac{m\pi}{2} - \frac{\pi}{4})} + O(R^{-1})$$
$$\text{as} \quad R \to \infty. \tag{B.77}$$

The imaginary part of ϕ_n^m is given explicitly by

$$\text{Im}\,\phi_n^m = \frac{2\pi(-1)^n}{(n-m)!} K^{n+1}\, e^{K(z+\zeta)}\, J_m(KR). \tag{B.78}$$

Alternatively, if $R > 0$,

$$\phi_n^m = \frac{P_n^m(\cos\theta)}{r^{n+1}} + \frac{2(-1)^n}{\pi(n-m)!} \int_0^\infty \frac{\mu^n K_m(\mu R)}{\mu^2 + K^2} g_{n-m}(\mu)\, \mathrm{d}\mu$$

$$+ \frac{2\pi i(-1)^n}{(n-m)!} K^{n+1}\, e^{K(z+\zeta)}\, H_m^{(1)}(KR) \tag{B.79}$$

$$= \frac{4}{\pi(n-m)!} \int_0^\infty \frac{\mu^n K_m(\mu R)}{\mu^2 + K^2} (\mu \cos\mu z + K \sin\mu z) f_{n-m}(\mu)\mathrm{d}\mu$$

$$+ \frac{2\pi i(-1)^n}{(n-m)!} K^{n+1}\, e^{K(z+\zeta)}\, H_m^{(1)}(KR), \tag{B.80}$$

where $g_n(\mu)$ and $f_n(\mu)$ are given by (B.22) and (B.23), respectively. A power series expansion is obtained from (B.75):

$$\phi_n^m = \frac{P_n^m(\cos\theta)}{r^{n+1}} + \sum_{s=m}^{\infty} A_{ns}^m r^s P_s^m(\cos\theta), \qquad r < 2|\zeta|, \tag{B.81}$$

where

$$A_{ns}^m = \frac{(-1)^{m+n+s}}{(n-m)!(s+m)!} \int_0^{\infty} \frac{\mu+K}{\mu-K} \mu^{n+s} e^{2\mu\zeta} \, d\mu. \tag{B.82}$$

Wave-free potentials

From (B.76) it is clear that the combination

$$\phi_{n+1}^m + \frac{K}{n-m+1}\phi_n^m, \quad n = 1, 2, 3, \ldots$$

corresponds to a wave-free singularity. Thus

$$\psi_n^m = \frac{P_{n+1}^m(\cos\theta)}{r^{n+2}} + \frac{K}{n-m+1}\frac{P_n^m(\cos\theta)}{r^{n+1}}$$
$$+ \frac{P_{n+1}^m(\cos\theta_1)}{r_1^{n+2}} - \frac{K}{n-m+1}\frac{P_n^m(\cos\theta_1)}{r_1^{n+1}} \tag{B.83}$$

represents a wave-free potential. This formula is given incorrectly in Wehausen and Laitone (1960, eqn 13.21). When the singularity is in the free surface these potentials reduce to zero unless $n+m$ is odd. In that case we obtain two distinct sets of wave-free potentials:

$$\psi_{2n}^{2m} = \frac{P_{2n}^{2m}(\cos\theta)}{r^{2n+1}} + \frac{K}{2n-2m}\frac{P_{2n-1}^{2m}(\cos\theta)}{r^{2n}}, \tag{B.84}$$

$$\psi_{2n+1}^{2m+1} = \frac{P_{2n+1}^{2m+1}(\cos\theta)}{r^{2n+2}} + \frac{K}{2n-2m}\frac{P_{2n}^{2m+1}(\cos\theta)}{r^{2n+1}}. \tag{B.85}$$

B.4 Three dimensions, finite depth

In this section we consider solutions of Laplace's equation in three dimensions which are singular at $x = \xi$, $y = \eta$, $z = \zeta \leq 0$ and which satisfy the free-surface boundary condition (1.13) and the bed boundary condition (2.1). The notation is as in §B.3 together with

$$r_2 = [(x - \xi)^2 + (y - \eta^2) + (z + \zeta + 2h)^2]^{1/2},$$
$$z_> = \max(z, \zeta), \qquad z_< = \min(z, \zeta).$$

The definitions of k and k_m, N_m, $m = 0, 1, 2, \ldots$ are those of §2.1.

Submerged source

A submerged source situated at (ξ, η, ζ), $\zeta < 0$, can be written

$$
\phi = \frac{1}{r} + \frac{1}{r_2} + \fint_0^\infty \frac{2(\mu + K) \, e^{-\mu h}}{\mu \sinh \mu h - K \cosh \mu h}
$$
$$
\times \cosh \mu(z + h) \cosh \mu(\zeta + h) J_0(\mu R) \, d\mu \qquad \text{(B.86)}
$$
$$
= -2 \fint_0^\infty \frac{K \sinh \mu z_> + \mu \cosh \mu z_>}{\mu \sinh \mu h - K \cosh \mu h} \cosh \mu(z_< + h) J_0(\mu R) \, d\mu \qquad \text{(B.87)}
$$
$$
\sim \frac{\pi i}{h N_0^2} \cosh k(z + h) \cosh k(\zeta + h) H_0^{(1)}(kR) \quad \text{as} \quad R \to \infty. \qquad \text{(B.88)}
$$

The imaginary part of ϕ is given explicitly by

$$
\operatorname{Im} \phi = \frac{\pi}{h N_0^2} \cosh k(z + h) \cosh k(\zeta + h) J_0(kR). \qquad \text{(B.89)}
$$

Provided $R > 0$, we can use the eigenfunction expansion

$$
\phi = \frac{\pi i}{h N_0^2} \cosh k(z + h) \cosh k(\zeta + h) H_0^{(1)}(kR)
$$
$$
+ \sum_{m=1}^\infty \frac{2}{h N_m^2} \cos k_m(z + h) \cos k_m(\zeta + h) K_0(k_m R) \qquad \text{(B.90)}
$$
$$
= \sum_{m=0}^\infty \frac{2}{h N_m^2} \cos k_m(z + h) \cos k_m(\zeta + h) K_0(k_m R) \qquad \text{(B.91)}
$$

since $K_0(k_0 R) = K_0(-ikR) = \frac{\pi i}{2} H_0^{(1)}(kR)$.

Multipoles

Multipoles singular at (ξ, η, ζ), $\zeta < 0$, are of the form $\phi_n^m \cos m\alpha$ and $\phi_n^m \sin m\alpha$, where

$$
\phi_n^m = \frac{P_n^m(\cos \theta)}{r^{n+1}} + \frac{1}{(n-m)!} \fint_0^\infty \frac{\mu^n J_m(\mu R) f_n^m(\mu)}{\mu \sinh \mu h - K \cosh \mu h} \, d\mu \qquad \text{(B.92)}
$$

with

$$f_n^m(\mu) = (-1)^m e^{-\mu(\zeta+h)}(K \sinh \mu z + \mu \cosh \mu z)$$
$$+ (-1)^n(\mu + K) e^{\mu\zeta} \cosh \mu(z+h). \tag{B.93}$$

The imaginary part of ϕ_n^m is given explicitly by

$$\operatorname{Im} \phi_n^m = \frac{\pi k^n \cosh k(z+h)}{2h N_0^2(n-m)!} \left((-1)^m e^{-k(\zeta+h)} + (-1)^n e^{k(\zeta+h)} \right) J_m(kR) \tag{B.94}$$

and as $R \to \infty$

$$\phi_n^m \sim \frac{\pi i k^n \cosh k(z+h)}{2h N_0^2(n-m)!} \left((-1)^m e^{-k(\zeta+h)} + (-1)^n e^{k(\zeta+h)} \right) H_m^{(1)}(kR). \tag{B.95}$$

A power series expansion can be obtained from (B.92):

$$\phi_n^m = \frac{P_n^m(\cos\theta)}{r^{n+1}} + \sum_{s=m}^{\infty} A_{ns}^m r^s P_s^m(\cos\theta), \qquad r < 2|\zeta|, \tag{B.96}$$

where

$$A_{ns}^m = \frac{1}{2(n-m)!(s+m)!} \int_0^\infty \frac{\mu^{n+s} g_{ns}^m(\mu)}{\mu \sinh \mu h - K \cosh \mu h} \, d\mu \tag{B.97}$$

with

$$g_{ns}^m(\mu) = (\mu + K) \left((-1)^{m+n+s} e^{\mu(2\zeta+h)} + ((-1)^s + (-1)^n) e^{-\mu h} \right)$$
$$+ (-1)^m(\mu - K) e^{-\mu(2\zeta+h)}. \tag{B.98}$$

Wave-free potentials

Wave-free singularities can be constructed as for the infinite depth case. Equation (B.95) shows that if $m+n$ is even the potentials

$$\phi_{n+1}^m + \frac{k}{n-m+1} \tanh k(\zeta + h) \phi_n^m$$

are wave free, whereas if $m+n$ is odd

$$\phi_{n+1}^m + \frac{k}{n-m+1} \coth k(\zeta + h) \phi_n^m$$

are wave free.

B.5   Oblique waves in infinite depth

In this section we consider solutions of the modified Helmholtz equation which are singular at $x = \xi$, $z = \zeta \leq 0$ and which satisfy the free-surface boundary condition (1.13). The coordinates that are used are as in §B.1. Throughout this section we will write

$$\nu = \ell \cosh \mu \qquad (B.99)$$

for notational convenience. Solutions to $(\nabla^2 - \ell^2)\phi(r, \theta) = 0$, which are singular at $r = 0$ are $K_n(\ell r) \cos n\theta$ and $K_n(\ell r) \sin n\theta$, where K_n is an nth order modified Bessel function. The behaviour of these functions for small arguments is given by (A.14) when $n = 0$ and by (A.15) for $n = 1, 2, 3, \ldots$. We have the following integral representations (see e.g. Ursell 1951):

$$K_n(\ell r) \cos n\theta = \int_0^\infty \cosh n\mu \cos(\ell X \sinh \mu) \, e^{\nu(z-\zeta)} \, d\mu, \quad z < \zeta, \quad (B.100)$$

$$= (-1)^n \int_0^\infty \cosh n\mu \cos(\ell X \sinh \mu) \, e^{\nu(\zeta-z)} \, d\mu, \quad z > \zeta, \quad (B.101)$$

$$K_n(\ell r) \sin n\theta = \int_0^\infty \sinh n\mu \sin(\ell X \sinh \mu) \, e^{\nu(z-\zeta)} \, d\mu, \quad z < \zeta, \quad (B.102)$$

$$= (-1)^{n+1} \int_0^\infty \sinh n\mu \sin(\ell X \sinh \mu) \, e^{\nu(\zeta-z)} \, d\mu, \quad z > \zeta. \quad (B.103)$$

Submerged source

A submerged source situated at (ξ, ζ), $\zeta < 0$, can be written

$$\phi = K_0(\ell r) - K_0(\ell r_1) + \fint_0^\infty \frac{2\nu}{\nu - K} \cos(\ell X \sinh \mu) \, e^{\nu(z+\zeta)} \, d\mu \quad (B.104)$$

$$\sim 2\pi i K \alpha^{-1} \, e^{K(z+\zeta)} \, e^{\pm i\alpha X} \quad \text{as} \quad R \to \infty, \quad (B.105)$$

where $\alpha = (K^2 - \ell^2)^{1/2}$. The imaginary part of ϕ is given explicitly by (B.114) with $n = 0$ and a power series expansion is given by (B.115), (B.116) with $n = 0$.

Free-surface source

A source situated at $(\xi, 0)$ can be written

$$\phi = \int_0^\infty \frac{\nu}{\nu - K} \cos(\ell X \sinh \mu) \, e^{\nu z} \, d\mu \tag{B.106}$$

$$\sim \pi i K \alpha^{-1} e^{Kz} e^{\pm i\alpha X} \quad \text{as} \quad R \to \infty. \tag{B.107}$$

The imaginary part of ϕ is given explicitly by

$$\text{Im}\, \phi = \pi K \alpha^{-1} e^{Kz} \cos \alpha X. \tag{B.108}$$

A power series expansion expansion can be obtained from (B.106) as shown in Ursell (1962, 2001):

$$\phi = K_0(\ell r) + (\pi i - \gamma) \coth \gamma \sum_{n=0}^\infty \epsilon_n (-1)^n I_n(\ell r) \cos n\theta \cosh n\gamma$$

$$- 2 \coth \gamma \sum_{n=1}^\infty (-1)^n \frac{\partial}{\partial \nu} \{I_\nu(\ell r) \cos \nu\theta\}_{\nu=n} \sinh n\gamma \tag{B.109}$$

in which ϵ_n is the Neumann symbol defined on page 17 and γ is defined by the equation

$$K = \ell \cosh \gamma. \tag{B.110}$$

If $K > \ell$, γ is taken to be positive whereas if $K < \ell$ then $\gamma = i\gamma^*$, where γ^* is positive.

Symmetric multipoles

Multipoles which are singular at (ξ, ζ), $\zeta < 0$, and symmetric about $x = \xi$ can be written

$$\phi_n = K_n(\ell r) \cos n\theta$$

$$+ (-1)^n \int_0^\infty A(\mu) \cosh n\mu \cos(\ell X \sinh \mu) \, e^{\nu(z+\zeta)} \, d\mu \tag{B.111}$$

$$\sim 2\pi i (-1)^n K \alpha^{-1} e^{K(z+\zeta)} \cosh n\gamma \, e^{\pm i\alpha X} \quad \text{as} \quad x \to \pm\infty, \tag{B.112}$$

where $\alpha = (K^2 - \ell^2)^{1/2}$ and

$$A(\mu) = \frac{\nu + K}{\nu - K}. \tag{B.113}$$

If $\ell > K$ these multipoles are real but if $\ell < K$ the path of integration is indented to pass beneath the pole at $\mu = \gamma$ (defined in equation B.110) and the imaginary part of ϕ_n is given explicitly by

$$\text{Im}\, \phi_n = 2\pi(-1)^n K\alpha^{-1}\, e^{K(z+\zeta)} \cosh n\gamma \cos \alpha X. \tag{B.114}$$

A power series expansion can be obtained from (B.111):

$$\phi_n = K_n(\ell r) \cos n\theta + \sum_{m=0}^{\infty} A_{mn} I_m(\ell r) \cos m\theta, \tag{B.115}$$

where

$$A_{mn} = \epsilon_m (-1)^{m+n} \int_0^\infty \cosh m\mu \cosh n\mu\, e^{2\nu\zeta}\, A(\mu)\, d\mu. \tag{B.116}$$

Antisymmetric multipoles

Multipoles which are singular at (ξ, ζ), $\zeta < 0$, and antisymmetric about $x = \xi$ can be written

$$\phi_n = K_n(\ell r) \sin n\theta$$
$$- (-1)^n \int_0^\infty A(\mu) \sinh n\mu \sin(\ell X \sinh \mu)\, e^{\nu(z+\zeta)}\, d\mu. \tag{B.117}$$
$$\sim \mp 2\pi(-1)^n K\alpha^{-1}\, e^{K(z+\zeta)} \sinh n\gamma\, e^{\pm i\alpha X} \quad \text{as} \quad x \to \pm\infty, \tag{B.118}$$

where $A(\mu)$ is again given by (B.113). These functions are real if $\ell > K$ and if $\ell < K$ the imaginary part of ϕ_n is given explicitly by

$$\text{Im}\, \phi_n = -2\pi(-1)^n K\alpha^{-1}\, e^{K(z+\zeta)} \sinh n\gamma \sin \alpha X. \tag{B.119}$$

A power series expansion can be obtained from (B.117):

$$\phi_n = K_n(\ell r) \sin n\theta + \sum_{m=1}^{\infty} A_{mn} I_m(\ell r) \sin m\theta, \tag{B.120}$$

where

$$A_{mn} = 2(-1)^{m+n} \int_0^\infty \sinh m\mu \sinh n\mu\, e^{2\nu\zeta}\, A(\mu)\, d\mu. \tag{B.121}$$

Wave-free potentials

It is clear from (B.112) that the combination of symmetric multipoles given by

$$\psi_n^s = \phi_{2n-2} + 2K\ell^{-1}\phi_{2n-1} + \phi_{2n} \tag{B.122}$$
$$= K_{2n-2}(\ell r)\cos(2n-2)\theta + 2K\ell^{-1}K_{2n-1}(\ell r)\cos(2n-1)\theta$$
$$+ K_{2n}(\ell r)\cos 2n\theta, \tag{B.123}$$

$n = 1, 2, 3, \ldots$, corresponds to a wave-free singularity, whereas from (B.118) it follows that the combination of antisymmetric multipoles

$$\psi_n^a = \phi_{2n-1} + 2K\ell^{-1}\phi_{2n} + \phi_{2n+1} \tag{B.124}$$
$$= K_{2n-1}(\ell r)\sin(2n-1)\theta + 2K\ell^{-1}K_{2n}(\ell r)\sin 2n\theta$$
$$+ K_{2n+1}(\ell r)\sin(2n+1)\theta, \tag{B.125}$$

$n = 1, 2, 3, \ldots$, is wave free.

B.6 Oblique waves in finite depth

In this section we consider solutions of the modified Helmholtz equation which are singular at $x = \xi$, $z = \zeta \leq 0$ and which satisfy both the free-surface boundary condition (1.13) and the bed boundary condition (2.1). The coordinates that are used are as in §B.2 and ν is again defined by (B.99). The definitions of k and N_0 are those of §2.1.

Symmetric multipoles

Multipoles which are singular at (ξ, ζ), $\zeta < 0$, and symmetric about $x = \xi$ can be written

$$\phi_n = K_n(\ell r)\cos n\theta + \int_0^\infty \frac{\cosh n\mu \cos(\ell X \sinh \mu)}{\nu \sinh \nu h - K \cosh \nu h} f_n(\nu) \, d\mu, \tag{B.126}$$

where f_n is defined in (B.50). If $\ell > k$ these multipoles are real but if $\ell < k$ the path of integration is indented to pass beneath the pole at

$\mu = \gamma = \cosh^{-1}(k/\ell)$ and the imaginary part of ϕ_n is given explicitly by

$$\mathrm{Im}\,\phi_n = \frac{\pi\left(\mathrm{e}^{-k(\zeta+h)} +(-1)^n\,\mathrm{e}^{k(\zeta+h)}\right)}{2\alpha h N_0^2}\cosh n\gamma\cosh k(z+h)\cos\alpha X,$$

(B.127)

where $\alpha = (k^2 - \ell^2)^{1/2}$. As $x \to \pm\infty$

$$\phi_n \sim \frac{\pi\mathrm{i}\left(\mathrm{e}^{-k(\zeta+h)} +(-1)^n\,\mathrm{e}^{k(\zeta+h)}\right)}{2\alpha h N_0^2}\cosh n\gamma\cosh k(z+h)\,\mathrm{e}^{\pm\mathrm{i}\alpha X}\,.$$

(B.128)

A power series expansion can be obtained from (B.126) and we find that ϕ_n can be written as in (B.115) with

$$A_{mn} = \frac{\epsilon_m}{2}\int_0^\infty \frac{\cosh m\mu\,\cosh n\mu}{\nu\sinh\nu h - K\cosh\nu h}g_{mn}(\nu)\,\mathrm{d}\mu,$$

(B.129)

where ϵ_m is defined on page 17 and g_{mn} is given by (B.55).

Antisymmetric multipoles

Multipoles which are singular at (ξ,ζ), $\zeta < 0$, and antisymmetric about $x = \xi$ can be written

$$\phi_n = K_n(\ell r)\sin n\theta + \int_0^\infty \frac{\sinh n\mu\sin\left(\ell X\sinh\mu\right)}{\nu\sinh\nu h - K\cosh\nu h}f_n(\nu)\,\mathrm{d}\mu,\quad\text{(B.130)}$$

where f_n is now defined by (B.57). These functions are real if $\ell > k$ and if $\ell < k$ the imaginary part of ϕ_n is given explicitly by

$$\mathrm{Im}\,\phi_n = \frac{\pi\left(\mathrm{e}^{-k(\zeta+h)} -(-1)^n\,\mathrm{e}^{k(\zeta+h)}\right)}{2\alpha h N_0^2}\sinh n\gamma\cosh k(z+h)\sin\alpha X,$$

(B.131)

where $\alpha = (k^2 - \ell^2)^{1/2}$. As $x \to \pm\infty$

$$\phi_n \sim \pm\frac{\pi\left(\mathrm{e}^{-k(\zeta+h)} -(-1)^n\,\mathrm{e}^{k(\zeta+h)}\right)}{2\alpha h N_0^2}\sinh n\gamma\cosh k(z+h)\,\mathrm{e}^{\pm\mathrm{i}\alpha X}\,.$$

(B.132)

A power series expansion can be obtained from (B.130) and we find that ϕ_n can be written as in (B.120) with

$$A_{mn} = \int_0^\infty \frac{\sinh m\mu\,\sinh n\mu}{\nu\sinh\nu h - K\cosh\nu h}g_{mn}(\nu)\,\mathrm{d}\mu,$$

(B.133)

where g_{mn} is given by (B.62).

Appendix C

Principal-value and finite-part integrals

For the interval $[a, b]$, denote by $C^n[a, b]$ the space of functions with n continuous derivatives and by $C^{n,\alpha}[a, b]$, $\alpha \in (0, 1]$, the space of functions with n Hölder continuous derivatives.

For a function $f \in C^{0,\alpha}[a, b]$ a Cauchy principal-value integral is defined by

$$\fint_a^b \frac{f(t)}{s-t}\,dt = \lim_{\epsilon \to 0} \left\{ \int_a^{s-\epsilon} \frac{f(t)}{s-t}\,dt + \int_{s+\epsilon}^b \frac{f(t)}{s-t}\,dt \right\}. \tag{C.1}$$

For $f \in C^1[a, b]$, integration by parts gives the regularization

$$\fint_a^b \frac{f(t)}{s-t}\,dt = f(a)\ln(s-a) - f(b)\ln(b-s) + \int_a^b f'(t)\ln|s-t|\,dt. \tag{C.2}$$

Further for $f \in C^{1,\alpha}[a, b]$, it follows from (C.2) that

$$\frac{d}{ds}\fint_a^b \frac{f(t)}{s-t}\,dt = \frac{f(a)}{s-a} + \frac{f(b)}{b-s} + \fint_a^b \frac{f'(t)}{s-t}\,dt. \tag{C.3}$$

For $f \in C^{1,\alpha}[a, b]$ a two-sided Hadamard finite-part integral of order two is defined by

$$\fint_a^b \frac{f(t)}{(s-t)^2}\,dt = \lim_{\epsilon \to 0} \left\{ \int_a^{s-\epsilon} \frac{f(t)}{(s-t)^2}\,dt + \int_{s+\epsilon}^b \frac{f(t)}{(s-t)^2}\,dt - \frac{2f(s)}{\epsilon} \right\} \tag{C.4}$$

which has the regularization

$$\oint_a^b \frac{f(t)}{(s-t)^2}\, dt = -\frac{f(a)}{s-a} - \frac{f(b)}{b-s} - \mathcal{P}\!\!\int_a^b \frac{f'(t)}{s-t}\, dt. \tag{C.5}$$

Comparison of (C.3) and (C.5) gives

$$\frac{d}{ds} \mathcal{P}\!\!\int_a^b \frac{f(t)}{s-t}\, dt = -\oint_a^b \frac{f(t)}{(s-t)^2}\, dt. \tag{C.6}$$

References

Abrahams, I. D. (1982). Scattering of sound by large finite geometries. *IMA J. Appl. Math. 29*, 79–97.

Abramowitz, M. and I. A. Stegun (1965). *Handbook of Mathematical Functions*. Dover Publications, New York.

Abul-Azm, A. G. and M. R. Gesraha (2000). Approximation to the hydrodynamics of floating pontoons under oblique waves. *Ocean Engng. 27*, 365–384.

Abul-Azm, A. G. and A. N. Williams (1989). Second-order diffraction loads on arrays of semi-immersed circular cylinders. *J. Fluids and Structures 3*, 365–387.

Achenbach, J. D. and Z. L. Li (1986). Reflection and transmission of scalar waves by a periodic array of screens. *Wave Motion 8*, 225–234.

Achenbach, J. D., Y.-C. Lu, and M. Kitahara (1988). 3-D reflection and transmission of sound by an array of rods. *J. Sound Vib. 125*(3), 463–476.

Alker, G. (1977). The scattering of short surface waves by a cylinder. *J. Fluid Mech. 82*, 673–686.

Aranha, J. A., C. C. Mei, and D. K. P. Yue (1979). Some properties of a hybrid element method for water waves. *Intl. J. Numer. Meth. Eng. 14*, 1627–1641.

Ashcroft, N. W. and N. D. Mermin (1976). *Solid State Physics*. W. B. Saunders, Philadelphia.

Athanassoulis, G. A. (1984). An expansion theorem for water-wave potentials. *J. Engng. Math. 18*, 181–194.

Bai, K. J. (1975). Diffraction of oblique waves by an infinite cylinder. *J. Fluid Mech. 68*, 513–535.

Baker, B. B. and E. T. Copson (1950). *The Mathematical Theory of Huygens' Principle* (2nd ed.). Clarendon Press, Oxford.

Balsa, T. F. (1982). Low frequency two dimensional flows through a sparse array of bodies. *J. Sound Vib. 82*(4), 489–504.

Balsa, T. F. (1983). Low frequency flows through an array of airfoils. *J. Sound Vib. 86*(3), 353–367.

Barakat, R. (1962). Vertical motion of a floating sphere in a sine-wave sea. *J. Fluid Mech. 13*, 540–556.

Barlow, H. E. M. and A. E. Karbowiak (1954). An experimental investigation of the properties of corrugated cylindrical surface waveguides. *Proc. IEE 101*(Part III), 182–188.

Batchelor, G. K. (1967). *An Introduction to Fluid Dynamics.* Cambridge University Press.

Berz, F. (1951). Reflection and refraction of microwaves at a set of parallel metallic plates. *Proc. IEE 98*(III), 47–55.

Biggs, N. R. T., D. Porter, and D. S. G. Stirling (2000). Wave diffraction through a perforated breakwater. *Q. J. Mech. Appl. Math. 53*, 375–391.

Birkhoff, G. and G.-C. Rota (1989). *Ordinary Differential Equations* (4th ed.). John Wiley & Sons, New York.

Black, J. L. (1975). Wave forces on vertical axisymmetric bodies. *J. Fluid Mech. 67*, 369–376.

Black, J. L., C. C. Mei, and C. G. Bray (1971). Radiation and scattering of water waves by rigid bodies. *J. Fluid Mech. 46*, 151–164.

Bolton, W. E. and F. Ursell (1973). The wave force on an infinitely long circular cylinder in an oblique sea. *J. Fluid Mech. 57*, 241–256.

Bonnet-Bendhia, A.-S. and F. Starling (1994). Guided waves by electromagnetic gratings and non-uniqueness examples for the diffraction problem. *Math. Meth. in the Appl. Sci. 17*, 305–338.

Bowman, J. J., T. B. A. Senior, and P. L. E. Uslenghi (Eds.) (1987). *Electromagnetic and Acoustic Scattering by Simple Shapes* (Revised ed.). Hemisphere, New York.

Burniston, E. E. and C. E. Siewert (1973). The use of Riemann problems in solving a class of transcendental equations. *Proc. Camb. Phil. Soc. 73*, 111–118.

Burrows, A. (1985). Waves incident on a circular harbour. *Proc. Roy. Soc. Lond., A 401*, 349–371.

Cadby, J. R. and C. M. Linton (2000). Three-dimensional water-wave scattering in two-layer fluids. *J. Fluid Mech. 423*, 155–173.

Carlson, J. F. and A. E. Heins (1947). The reflection of an electromagnetic plane wave by an infinite set of plates, I. *Quart. Appl. Math. 4*, 313–329.

Carr, J. H. and M. E. Stelzriede (1952). Diffraction of water waves by breakwaters. *US Nat. Bur. Stds. 521*, 109–125.

Chakrabarti, S. K. (2000). Hydrodynamic interaction forces on multi-moduled structures. *Ocean Engng. 27*, 1037–1063.

Chamberlain, P. G. and D. Porter (1999). On the solution of the dispersion relation for water waves. *Appl. Ocean Res. 21*, 161–166.

Chen, H. S. and C. C. Mei (1973). Wave forces on a stationary platform of elliptical shape. *J. Ship Res. 17*(2), 61–71.

Chou, T. (1998). Band structure of surface flexural-gravity waves along periodic interfaces. *J. Fluid Mech. 369*, 333–350.

Chwang, A. T. and J. Wu (1994). Wave scattering by submerged porous disk. *J. Engrg. Mech. 120*, 2575–2587.

Colton, D. and R. Kress (1983). *Integral Equation Methods in Scattering Theory.* John Wiley & Sons, New York.

Courant, R. and D. Hilbert (1953). *Methods of Mathematical Physics. Volume I.* Interscience Publishers, New York.

Crapper, G. D. (1984). *Introduction to Water Waves.* Ellis Horwood, Chichester.

Crighton, D. G., A. P. Dowling, J. E. Ffowcs Williams, M. Heckl, and F. G. Leppington (1992). *Modern Methods in Analytical Acoustics.* Springer-Verlag, London.

Crighton, D. G. and F. G. Leppington (1973). Singular perturbation methods in acoustics: diffraction by a plate of finite thickness. *Proc. Roy. Soc. Lond., A 335*, 313–339.

Curzon, F. L. and D. Plant (1986). Using perturbed resonant frequencies to study eigenmodes of an acoustic resonator. *Am. J. Phys. 54*, 367–372.

Dalrymple, R. A. and P. A. Martin (1988). Water wave scattering by rows of circular cylinders. In *Proc. 21st Int. Conference on Coastal Engineering, ASCE*, Torremolinos, pp. 2216–2228.

Dalrymple, R. A. and P. A. Martin (1990). Wave diffraction through offshore breakwaters. *J. Wtrwy., Port, Coast., and Oc. Engrg., ASCE 116*(6), 727–741.

Das, A. and M. S. Lundstrom (1990). A scattering matrix approach to device simulation. *Solid-St. Electron. 33*(10), 1299–1307.

Davidovitz, M. and Y. T. Lo (1987). Cutoff wavenumbers and modes for annular-cross-section waveguide with eccentric inner conductor of small radius. *IEEE Trans. Microwave Theory Tech. 35*(5), 510–515.

Davis, A. M. J. (1975). Small oscillations in a hemispherical lake. *Q. J. Mech. Appl. Math. 28*, 157–179.

Davis, A. M. J. (1976). A relation between the radiation and scattering of surface waves by axisymmetric bodies. *J. Fluid Mech. 76*, 85–88.

Davis, A. M. J. and F. G. Leppington (1977). The scattering of electromagnetic surface waves by circular or elliptic cylinders. *Proc. Roy. Soc. Lond., A 353*, 55–75.

Davis, A. M. J. and F. G. Leppington (1985). Scattering of long surface waves by circular cylinders or spheres. *Q. J. Mech. Appl. Math. 38*, 411–432.

Dawson, T. W. (1991). Scattering matrix and boundary integral equation methods for long-range propagation in an acoustic waveguide with repeated boundary deformations. *J. Acoust. Soc. Am. 90*(3), 1560–1581.

Dean, W. R. (1948). On the reflexion of surface waves by a submerged circular cylinder. *Proc. Camb. Phil. Soc. 44*, 483–491.

Delves, L. M. and J. L. Mohamed (1985). *Computational Methods for Integral Equations.* Cambridge University Press.

Dorfmann, A. A. and A. A. Savvin (1998). Diffraction of water waves by a horizontal plate. *J. Applied Math. Phys. (ZAMP) 49*, 805–826.

Drake, K. R. (1999). The effect of internal pipes on the fundamental frequency of liquid sloshing in a circular tank. *Appl. Ocean Res. 21*, 133–143.

Dudley, D. G. (1994). *Mathematical Foundations for Electromagnetic Theory.* IEEE, New York/Oxford University Press.

Duff, G. and D. Naylor (1966). *Differential Equations of Applied Mathematics.* John Wiley, New York.

Duncan, J. H. and C. E. Brown (1982). Development of a numerical method for the calculation of power absorption by arrays of similar arbitrarily shaped bodies in a seaway. *J. Ship Res. 26*, 38–44.

Eatock Taylor, R. and C. S. Hu (1991). Multipole expansions for wave diffraction and radiation in deep water. *Ocean Engng. 18*(3), 191–224.

Edwards, C. H. and D. E. Penney (1988). *Elementary Linear Algebra.* Prentice-Hall, Englewood Cliffs, New Jersey.

Endo, H. (1987). Shallow-water effect on the motions of three-dimensional bodies in waves. *J. Ship Res. 31*, 34–40.

Ervin, V. J. and E. P. Stephan (1992). Collocation with Chebyshev polynomials for a hypersingular integral equation on an interval. *J. Comput. Appl. Math. 43*, 221–229.

Esquivel-Sirvent, R. and G. H. Cocoletzi (1994). Band structure for the propagation of elastic waves in superlattices. *J. Acoust. Soc. Am. 95*(1), 86–90.

Estrada, R. and R. P. Kanwal (1989). Integral equations with logarithmic kernels. *IMA J. Appl. Math. 43*, 133–155.

Evans, D. V. (1980). Some analytic results for two- and three-dimensional wave-energy absorbers. In B. M. Count (Ed.), *Power from Sea Waves.* Academic Press.

Evans, D. V. (1990). The wide spacing approximation applied to multiple scattering and sloshing problems. *J. Fluid Mech. 210*, 647–658.

Evans, D. V. (1992). Trapped acoustic modes. *IMA J. Appl. Math. 49*(1), 45–60.

Evans, D. V. and M. Fernyhough (1995). Edge waves along periodic coastlines. Part 2. *J. Fluid Mech. 297*, 301–325.

Evans, D. V., D. C. Jeffrey, S. H. Salter, and J. R. M. Taylor (1979). The submerged cylinder wave energy device: theory and experiment. *Appl. Ocean Res. 1*, 3–12.

Evans, D. V. and C. M. Linton (1989). Active devices for the reduction in wave intensity. *Appl. Ocean Res. 11*(1), 26–32.

Evans, D. V. and C. M. Linton (1993a). Edge waves along periodic coastlines. *Q. J. Mech. Appl. Math. 46*(4), 642–656.

Evans, D. V. and C. M. Linton (1993b). Sloshing frequencies. *Q. J. Mech. Appl. Math. 46*(1), 71–87.

Evans, D. V. and C. M. Linton (1994). On step approximations for water wave problems. *J. Fluid Mech. 278*, 229–249.

Evans, D. V. and P. McIver (1987). Resonant frequencies in a container with a vertical baffle. *J. Fluid Mech. 175*, 295–307.

Evans, D. V. and C. A. N. Morris (1972). Complementary approximations to the solution of a problem in water waves. *J. Inst. Maths Applics. 10*, 1–9.

Evans, D. V. and R. Porter (1996). Hydrodynamic characteristics of a thin rolling plate in finite depth of water. *Appl. Ocean Res. 18*(4), 215–228.

Falnes, J. (1980). Radiation impedance matrix and optimum power absorption for interacting oscillators in surface waves. *Appl. Ocean Res. 2*, 75–80.

Falnes, J. and P. McIver (1985). Surface wave interactions with systems of oscillating bodies and pressure distributions. *Appl. Ocean Res. 7*, 225–234.

Faltinsen, O. M. (1990). *Sea Loads on Ships and Offshore Structures.* Cambridge University Press.

Farell, C. (1973). On the wave resistance of a submerged spheroid. *J. Ship Res. 17*, 1–11.

Farina, L. and P. A. Martin (1998). Scattering of water waves by a submerged disc using a hypersingular integral equation. *Appl. Ocean Res. 20*(3), 121–134.

Fenton, J. D. (1978). Wave forces on vertical bodies of revolution. *J. Fluid Mech. 85*, 241–255.

Fox, D. W. and J. R. Kuttler (1981). Upper and lower bounds for sloshing frequencies by intermediate problems. *J. Applied Math. Phys. (ZAMP) 32*, 667–682.

Fox, D. W. and J. R. Kuttler (1983). Sloshing frequencies. *J. Applied Math. Phys. (ZAMP) 34*, 668–696.

Frenkel, A. (1983). External modes of two-dimensional thin scatterers. *IEE Proc. H, Microwaves, Opt. & Antennas 130*(3), 209–214.

Friedrichs, K. O. and H. Lewy (1948). The dock problem. *Comm. Applied Maths 1*, 135–148.

Garrett, C. J. R. (1971). Wave forces on a circular dock. *J. Fluid Mech. 46*, 129–139.

Garrison, C. J. (1969). On the interaction of an infinite shallow draft cylinder oscillating at the free surface with a train of oblique waves. *J. Fluid Mech. 39*, 227–255.

Garrison, C. J. (1985). Interaction of oblique waves with an infinite cylinder. In C. A. Brebbia (Ed.), *Boundary Element Research*, pp. 32–43. CML Publications, Southampton.

Gilbert, G. and A. H. Brampton (1985). The solution of two wave diffraction problems using integral equations. Technical Report IT 299, Hydraulics Research, Wallingford.

Golberg, M. A. (1983). The convergence of several algorithms for solving integral equations with finite part integrals. *J. Integral Equations 5*, 329–340.

Golberg, M. A. (1985). The convergence of several algorithms for solving integral equations with finite part integrals. II. *J. Integral Equations 9*, 267–275.

Gradshteyn, I. S. and I. M. Ryzhik (1980). *Tables of Integrals, Series and Products* (4th ed.). Academic Press, New York.

Graham, E. W. and A. M. Rodriguez (1952). The characteristics of fuel motion which affect airplane dynamics. *J. Appl. Mech. 19*, 381–388.

Gray, E. P. (1978). Scattering of a surface wave by a submerged sphere. *J. Engng. Math. 12*(1), 15–41.

Green, M. W. (1971). A problem connected with the oblique incidence of surface waves on an immersed cylinder. *J. Inst. Maths Applics. 8*, 82–98.

Greene, T. H. and A. E. Heins (1953). Water waves over a channel of infinite depth. *Quart. Appl. Math. 11*, 201–214.

Greenhow, M. J. L. (1980). The hydrodynamic interactions of spherical wave power devices in surface waves. In B. M. Count (Ed.), *Power from Sea Waves*, pp. 287–343. Academic Press.

Guiggiani, M., G. Krishnasamy, T. J. Rudolphi, and F. J. Rizzo (1992). A general algorithm for the numerical solution of hypersingular boundary integral equations. *J. Appl. Mech. 59*, 604–614.

Havelock, T. H. (1929). Forced surface-waves on water. *Phil. Mag. 8*, 569–576.

Havelock, T. H. (1955). Waves due to a floating sphere making periodic heaving oscillations. *Proc. Roy. Soc. Lond., A 231*, 1–7.

Hearn, G. E. (1977). Alternative methods of evaluating Green's function in three-dimensional ship-wave problems. *J. Ship Res. 21*, 89–93.

Heaviside, O. (1893). *Electromagnetic Theory.* "The Electrician" Printing & Publishing Co., London (Dover reprint 1950).

Heckl, M. A. (1992). Sound propagation in bundles of periodically arranged cylindrical tubes. *Acustica 77*, 143–152.

Heckl, M. A. and L. S. Mulholland (1995). Some recent developments in the theory of acoustic transmission in tube bundles. *J. Sound Vib. 179*(1), 37–62.

Heins, A. E. (1948). Water waves over a channel of finite depth with a dock. *Amer. J. Math. 70*, 730–748.

Heins, A. E. (1950a). The reflection of an electromagnetic plane wave by an infinite set of plates, III. *Quart. Appl. Math. 8*, 281–291.

Heins, A. E. (1950b). Water waves over a channel of finite depth with a submerged plane barrier. *Can. J. Math. 2*, 210–222.

Heins, A. E. and J. F. Carlson (1947). The reflection of an electromagnetic plane wave by an infinite set of plates, II. *Quart. Appl. Math. 5*, 82–88.

Henrici, P., B. A. Troesch, and L. Wuytack (1970). Sloshing frequencies for a half-space with circular or strip-like aperture. *J. Applied Math. Phys. (ZAMP) 21*, 285–317.

Holford, R. L. (1964a). Short surface waves in the presence of a finite dock. I. *Proc. Camb. Phil. Soc. 60*, 957–983.

Holford, R. L. (1964b). Short surface waves in the presence of a finite dock. II. *Proc. Camb. Phil. Soc. 60*, 985–1011.

Hudspeth, R. T., T. Nakamura, and C.-K. Pyun (1994). Convergence criteria for axisymmetric Green's function with application to floating bodies. *Ocean Engng. 21*, 381–400.

Hulme, A. (1982). The wave forces on a floating hemisphere undergoing forced periodic oscillations. *J. Fluid Mech. 121*, 443–463.

Hulme, A. (1983). A ring-source/integral-equation method for the calculation of hydrodynamic forces exerted on floating bodies of revolution. *J. Fluid Mech. 128*, 387–412.

Hulme, A. (1985). The heave added-mass and damping coefficients of a submerged torus. *J. Fluid Mech. 155*, 511–530.

Hunt, B. (1990). An integral equation solution for the breakwater gap problem. *J. Hydraulic Research 28*, 609–619.

Hurd, R. A. (1954). The propagation of an electromagnetic wave along an infinite corrugated surface. *Can. J. Phys. 32*, 727–734.

Isaacson, M., T. Mathai, and C. Mihelcic (1990). Hydrodynamic coefficients of a vertical circular cylinder. *Can. J. Civil Eng. 17*, 302–310.

Isaacson, M. and K. Subbiah (1991). Earthquake induced sloshing in a rigid circular tank. *Can. J. Civil Eng. 18*, 904–915.

Isaacson, M. d. S. Q. (1978). Vertical cylinders of arbitrary section in waves. *J. Wtrwy., Port, Coast., and Oc. Division, ASCE 104*, 309–324.

Isaacson, M. d. S. Q. (1982). Fixed and floating axisymmetric structures in waves. *J. Wtrwy., Port, Coast., and Oc. Engrg., ASCE 108*, 180–199.

Itoh, T. and R. Mittra (1969). An analytical study of echelette grating with application to open resonators. *IEEE Trans. Microwave Theory Tech. 17*(6), 319–327.

John, F. (1950). On the motion of floating bodies, II. *Comm. Pure Appl. Maths 3*, 45–101.

Johnson, R. S. (1997). *A Modern Introduction to the Mathematical Theory of Water Waves.* Cambridge Texts in Applied Mathematics. Cambridge University Press.

Jones, D. S. (1952). A simplifying technique in the solution of a class of diffraction problems. *Quarterly Journal of Mathematics 3*, 189–196.

Jones, D. S. (1953). The eigenvalues of $\nabla^2 u + \lambda u = 0$ when the boundary conditions are given on semi-infinite domains. *Proc. Camb. Phil. Soc. 49*, 668–684.

Jones, D. S. (1986). *Acoustic and Electromagnetic Waves.* Clarendon Press, Oxford.

Jones, D. S. (1994). *Methods in Electromagnetic Wave Propagation* (2nd ed.). Clarendon Press, Oxford.

Kagemoto, H. and D. K. P. Yue (1986). Interactions among multiple three-dimensional bodies in water waves: an exact algebraic method. *J. Fluid Mech. 166*, 189–209.

Karp, S. N. and A. Russek (1956). Diffraction by a wide slit. *J. Appl. Phys. 27*, 886–894.

Kashiwagi, M. (1999). A hierarchical interaction theory for wave forces on a great number of buoyancy bodies. In *Proc. 14th Intl. Workshop on Water Waves and Floating Bodies*, Port Huron, Michigan, USA, pp. 68–71.

Kellogg, O. D. (1953). *Foundations of Potential Theory*. Dover Publications, New York.

Kim, M.-H. (1993). Interaction of waves with n vertical circular cylinders. *J. Wtrwy., Port, Coast., and Oc. Engrg., ASCE 119*, 671–689.

Kim, M.-H. and D. K. P. Yue (1989). The complete second-order diffraction solution for an axisymmetric body. Part 1. Monochromatic incident waves. *J. Fluid Mech. 200*, 235–264.

Kim, W. D. (1965). On the harmonic oscillations of a rigid body on a free surface. *J. Fluid Mech. 21*, 427–451.

Kobayashi, K. (1991). Plane wave diffraction by a strip: Exact and asymptotic solutions. *J. Phys. Soc. Jpn. 60*, 1891–1905.

Koch, W. (1971). On the transmission of sound waves through a blade row. *J. Sound Vib. 18*, 111–128.

Kotik, J. and V. Mangulis (1962). On the Kramers-Kronig relations for ship motions. *Int. Shipbuilding Progress 9*, 3–10.

Krishnasamy, G., L. W. Schmer, T. J. Rudolphi, and F. J. Rizzo (1990). Hypersingular boundary integral equations: some applications in acoustic and elastic wave scattering. *J. Appl. Mech. 57*, 404–414.

Kuttler, J. R. and V. G. Sigillito (1978). Bounding eigenvalues of elliptic operators. *SIAM J. Math. Anal. 9*, 768–773.

Kuttler, J. R. and V. G. Sigillito (1984). Eigenvalues of the Laplacian in two dimensions. *SIAM Review 26*(2), 163–193.

Kyllingstad, A. (1984). A low-scattering approximation for the hydrodynamic interactions of small wave-power devices. *Appl. Ocean Res. 6*, 132–139.

Lamb, H. (1932). *Hydrodynamics* (6th ed.). Cambridge University Press.

Lau, S. M. and G. E. Hearn (1989). The suppression of irregular frequency effects in fluid-structure interaction problems using a combined boundary integral equation method. *Intl. J. Numer. Meth. Fluids 9*, 763–782.

Lavretsky, E. I. (1994). Taking into account the edge condition in the problem of scattering from the circular aperture in circular-to-rectangular and rectangular-to-rectangular waveguide junctions. *IEE Proc. Microw. Antennas Propag. 141*, 45–50.

Lawrence, H. R., C. J. Wang, and R. B. Reddy (1958). Variational solution of fuel sloshing modes. *Jet Propulsion 28*, 729–736.

Lee, C.-H., J. N. Newman, and X. Zhu (1996). An extended boundary integral equation method for the removal of irregular frequency effects. *Intl. J. Numer. Meth. Fluids 23*, 637–660.

Lee, C.-H. and P. D. Sclavounos (1989). Removing the irregular frequencies from integral equations in wave-body interactions. *J. Fluid Mech. 207*, 393–418.

Leppington, F. G. (1968). On the scattering of short surface waves by a finite dock. *Proc. Camb. Phil. Soc. 64*, 1109–1129.

Leppington, F. G. (1970). On the radiation of short surface waves by a finite dock. *J. Inst. Maths Applics. 6*, 319–340.

Leppington, F. G. (1972). On the radiation and scattering of short surface waves. Part 1. *J. Fluid Mech. 56*, 101–119.

Leppington, F. G. and P. F. Siew (1980). Scattering of surface waves by submerged cylinders. *Appl. Ocean Res. 2*, 129–137.

Lesser, M. B. and J. A. Lewis (1972a). Applications of matched asymptotic expansion methods to acoustics. I. The Webster horn equation and the stepped duct. *J. Acoust. Soc. Am. 51*, 1664–1669.

Lesser, M. B. and J. A. Lewis (1972b). Applications of matched asymptotic expansion methods to acoustics. II. The open-ended duct. *J. Acoust. Soc. Am. 52*, 1406–1410.

Lesser, M. B. and J. A. Lewis (1974). The acoustic cavity containing small scatterers as a singular perturbation problem. *J. Sound Vib. 33*, 13–27.

Levine, L. (1965). Scattering of surface waves by a submerged circular cylinder. *J. Math. Phys. 6*(8), 1231–1243.

Liapis, S. (1992). Numerical methods for water-wave radiation problems. *Intl. J. Numer. Meth. Fluids 15*, 83–97.

Linton, C. M. (1988). *Wave reflection by submerged bodies in water of finite depth.* Ph.D. thesis, University of Bristol.

Linton, C. M. (1991). Radiation and diffraction of water waves by a submerged sphere in finite depth. *Ocean Engng. 18*(1/2), 61–74.

Linton, C. M. (1995). Multipole methods for boundary-value problems involving a sphere in a tube. *IMA J. Appl. Math. 55*, 187–204.

Linton, C. M. (1997). The use of multipoles in channel problems. In B. N. Mandal (Ed.), *Mathematical Techniques for Water Waves*, International Series on Advances in Fluid Mechanics, Chapter 2, pp. 45–78. Computational Mechanics Publications, Southampton.

Linton, C. M. (1998). The Green's function for the two-dimensional Helmholtz equation in periodic domains. *J. Engng. Math. 33*, 377–402.

Linton, C. M. (1999a). A new representation for the free-surface channel Green's function. *Appl. Ocean Res. 21*, 17–25.

Linton, C. M. (1999b). Rapidly convergent representations for Green's functions for Laplace's equation. *Proc. Roy. Soc. Lond., A 455*, 1767–1797.

Linton, C. M. and D. V. Evans (1990). The interaction of waves with arrays of vertical circular cylinders. *J. Fluid Mech. 215*, 549–569.

Linton, C. M. and D. V. Evans (1991). Trapped modes above a submerged horizontal plate. *Q. J. Mech. Appl. Math. 44*(3), 487–506.

Linton, C. M. and D. V. Evans (1992). The radiation and scattering of surface waves by a vertical circular cylinder in a channel. *Phil. Trans. R. Soc. Lond., A 338*, 325–357.

Linton, C. M. and D. V. Evans (1993a). Acoustic scattering by an array of parallel plates. *Wave Motion 18*, 51–65.

Linton, C. M. and D. V. Evans (1993b). Hydrodynamic characteristics of bodies in channels. *J. Fluid Mech. 252*, 647–666.

Linton, C. M. and D. V. Evans (1993c). The interaction of waves with a row of circular cylinders. *J. Fluid Mech. 251*, 687–708.

Linton, C. M. and N. G. Kuznetsov (1997). Non-uniqueness in two-dimensional water-wave problems: numerical evidence and geometrical restrictions. *Proc. Roy. Soc. Lond., A 453*, 2437–2460.

Linton, C. M. and M. McIver (1995). The interaction of waves with horizontal cylinders in two-layer fluids. *J. Fluid Mech. 304*, 213–229.

Linton, C. M. and P. McIver (1996). The scattering of water waves by an array of circular cylinders in a channel. *J. Engng. Math. 30*, 661–682.

MacCamy, R. C. (1958a). On Babinet's principle. *Can. J. Math. 10*, 632–640.

MacCamy, R. C. (1958b). On singular integral equations with logarithmic or Cauchy kernels. *J. Math. Mech. 7*, 355–375.

MacCamy, R. C. (1961). On the heaving motion of cylinders of shallow draft. *J. Ship Res. 5*(3), 34–43.

MacCamy, R. C. and R. A. Fuchs (1954). Wave forces on piles: A diffraction theory. Tech. Mem. 69, US Army Coastal Engineering Research Center.

Mandal, B. N. and A. Chakrabarti (2000). *Water Wave Scattering by Barriers*. WIT Press, Southampton.

Martin, P. A. (1980). On the null-field equations for the exterior problems of acoustics. *Q. J. Mech. Appl. Math. 33*, 385–396.

Martin, P. A. (1981). On the null-field equations for water-wave radiation problems. *J. Fluid Mech. 113*, 315–332.

Martin, P. A. (1984). On the null-field equations for water-wave scattering problems. *IMA J. Appl. Math. 33*, 55–69.

Martin, P. A. (1985a). Multiple scattering of surface water waves and the null-field method. In *Proc. 15th Symp. Naval Hydrodynamics*, Hamburg, pp. 119–132.

Martin, P. A. (1985b). On the T-matrix for water-wave scattering problems. *Wave Motion 7*, 177–193.

Martin, P. A. (1989). On the computation and excitation of trapping modes. In *Proc. 4th Intl. Workshop on Water Waves and Floating Bodies*, Øystese, Norway, pp. 145–148.

Martin, P. A. (1991a). End-point behaviour of solutions to hypersingular integral equations. *Proc. Roy. Soc. Lond., A 432*, 301–320.

Martin, P. A. (1991b). Ursell's multipoles and the Rayleigh hypothesis. In *Proc. 6th Intl. Workshop on Water Waves and Floating Bodies*, Woods Hole, Massachusetts, pp. 163–167.

Martin, P. A. (1995). Asymptotic approximations for functions defined by series, with some applications in the theory of guided waves. *IMA J. Appl. Math. 54*, 139–157.

Martin, P. A. and R. A. Dalrymple (1988). Scattering of long waves by cylindrical obstacles and gratings using matched asymptotic expansions. *J. Fluid Mech. 188*, 465–490.

Martin, P. A. and L. Farina (1997). Radiation of water waves by a heaving submerged horizontal disc. *J. Fluid Mech. 337*, 365–379.

Martin, P. A. and F. J. Rizzo (1989). On boundary integral equations for crack problems. *Proc. Roy. Soc. Lond., A 421*, 341–355.

Martin, P. A. and F. J. Rizzo (1996). Hypersingular integral equations: how smooth must the density be? *Intl. J. Numer. Meth. Eng. 39*, 687–704.

Mattioli, F. (1980). Numerical instabilities of the integral approach to the interior boundary-value problem for the two-dimensional Helmholtz equation. *Intl. J. Numer. Meth. Eng. 15*, 1303–1313.

Mavrakos, S. A. (1991). Hydrodynamic coefficients for groups of interacting vertical axisymmetric bodies. *Ocean Engng. 18*, 485–515.

Mavrakos, S. A. and P. Koumoutsakos (1987). Hydrodynamic interaction among vertical axisymmetric bodies restrained in waves. *Appl. Ocean Res. 9*, 128–140.

McIver, M. (1985a). *The interaction of water waves with submerged bodies*. Ph.D. thesis, University of Bristol.

McIver, M. (1996a). An example of non-uniqueness in the two-dimensional linear water wave problem. *J. Fluid Mech. 315*, 257–266.

McIver, M. (1996b). Global relations between two-dimensional water wave potentials. *J. Fluid Mech. 312*, 299–309.

McIver, M. (1997). Resonance in the unbounded water wave problem. In *Proc. 12th Intl. Workshop on Water Waves and Floating Bodies*, Carry-le-Rouet, France.

McIver, M. (2000a). Trapped modes supported by submerged obstacles. *Proc. Roy. Soc. Lond., A 456*, 1851–1860.

McIver, P. (1985b). Scattering of water waves by two surface-piercing vertical barriers. *IMA J. Appl. Math. 35*, 339–355.

McIver, P. (1987). Mean drift forces on arrays of bodies due to incident long waves. *J. Fluid Mech. 185*, 469–482.

McIver, P. (1991). Trapping of surface water waves by fixed bodies in a channel. *Q. J. Mech. Appl. Math. 44*(2), 193–208.

McIver, P. (1994a). Low-frequency asymptotics of hydrodynamic forces on fixed and floating structures. In M. Rahman (Ed.), *Ocean Waves Engineering*, International Series on Advances in Fluid Mechanics. Computational Mechanics Publications, Southampton.

McIver, P. (1994b). Transient fluid motion due to the forced horizontal oscillations of a vertical cylinder. *Appl. Ocean Res. 16*(6), 347–351.

McIver, P. (1998). The dispersion relation and eigenfunction expansions for water waves in a porous structure. *J. Engng. Math. 34*, 319–334.

McIver, P. (2000b). Water-wave propagation through an infinite array of cylindrical structures. *J. Fluid Mech. 424*, 101–125.

McIver, P. and G. S. Bennett (1993). Scattering of water waves by axisymmetric bodies in a channel. *J. Engng. Math. 27*, 1–29.

McIver, P. and D. V. Evans (1984a). Approximation of wave forces on cylinder arrays. *Appl. Ocean Res. 6*(2), 101–107.

McIver, P. and D. V. Evans (1984b). The occurrence of negative added mass in free-surface problems involving submerged oscillating bodies. *J. Engng. Math. 18*, 7–22.

McIver, P. and D. V. Evans (1985). The trapping of surface waves above a submerged horizontal cylinder. *J. Fluid Mech. 151*, 243–255.

McIver, P. and C. M. Linton (1994). Mean drift forces on arrays of axisymmetric structures in a wave tank. *Appl. Ocean Res. 16*(6), 327–335.

McIver, P., C. M. Linton, and M. McIver (1998). Construction of trapped modes for wave guides and diffraction gratings. *Proc. Roy. Soc. Lond., A 454*, 2593–2616.

McIver, P. and M. McIver (1997). Trapped modes in an axisymmetric water wave problem. *Q. J. Mech. Appl. Math. 50*, 165–178.

McIver, P. and A. D. Rawlins (1992). Scattering of water waves by a pair of semi-infinite barriers. *J. Appl. Mech. 59*, 1023–1025.

McIver, P. and A. D. Rawlins (1993). Two-dimensional wave scattering problems involving parallel-walled ducts. *Q. J. Mech. Appl. Math. 46*, 89–116.

McLachlan, N. W. (1954). *Bessel Functions for Engineers* (2nd ed.). Clarendon Press, Oxford.

Mehl, J. B. and R. N. Hill (1989). Acoustic eigenfrequencies of cavities with an internal obstacle: A modified perturbation theory. *J. Acoust. Soc. Am. 85*(5), 1841–1851.

Mei, C. C. (1978). Numerical methods in water-wave diffraction and radiation. *Ann. Rev. Fluid Mech. 10*, 393–416.

Mei, C. C. (1983). *The Applied Dynamics of Ocean Surface Waves.* Wiley-Interscience, New York.

Mei, C. C. and J. L. Black (1969). Scattering of surface waves by rectangular obstacles in waters of finite depth. *J. Fluid Mech. 38*, 499–511.

Miles, J. W. (1967). Surface-wave scattering matrix for a shelf. *J. Fluid Mech. 28*, 755–767.

Miles, J. W. (1971). A note on variational principles for surface-wave scattering. *J. Fluid Mech. 46*, 141–149.

Miles, J. W. (1982). On Rayleigh scattering by a grating. *Wave Motion 4*, 285–292.

Miles, J. W. (1983). Surface-wave diffraction by a periodic row of submerged ducts. *J. Fluid Mech. 128*, 155–180.

Miles, J. W. (1987). On surface wave forcing by a circular disk. *J. Fluid Mech. 175*, 97–108.

Miles, J. W. and F. Gilbert (1968). Scattering of gravity waves by a circular dock. *J. Fluid Mech. 34*, 783–793.

Milne-Thomson, L. M. (1996). *Theoretical Hydrodynamics* (5th ed.). Dover Publications, New York.

Mittra, R. and S. W. Lee (1971). *Analytical Techniques in the Theory of Guided Waves.* Macmillan, New York.

Moiseev, N. N. (1964). Introduction to the theory of oscillations of liquid-containing bodies. *Adv. Appl. Mech. 8*, 233–289.

Moiseev, N. N. and A. A. Petrov (1965). The calculation of free oscillations of a liquid in a motionless container. *Adv. Appl. Mech. 9*, 91–154.

Monacella, V. J. (1966). The distribution due to a slender ship oscillating in a fluid of finite depth. *J. Ship Res. 10*, 242–252.

Morse, P. M. and P. J. Rubinstein (1938). The diffraction of waves by ribbons and slits. *Phys. Rev. 54*, 895–898.

Mulholland, L. S. and M. A. Heckl (1994). Multi-directional sound-wave propagation through a tube bundle. *J. Sound Vib.* *176*(3), 377–398.

Naftzger, R. A. and S. K. Chakrabarti (1979). Scattering of waves by two-dimensional circular obstacles in finite water depths. *J. Ship Res.* *23*, 32–42.

Nestegard, A. and P. D. Sclavounos (1984). A numerical solution of two-dimensional deep water wave-body problems. *J. Ship Res.* *28*, 48–54.

Newman, J. N. (1965). Propagation of water waves past a long two-dimensional obstacle. *J. Fluid Mech.* *23*, 23–29.

Newman, J. N. (1976). The interaction of stationary vessels with regular waves. In *Proc. 11th Symp. on Naval Hydrodynamics*, London, pp. 491–501.

Newman, J. N. (1977). *Marine Hydrodynamics.* MIT Press, Cambridge, Massachusetts.

Newman, J. N. (1984a). Approximations for the Bessel and Struve functions. *Math. Comp.* *43*, 551–556.

Newman, J. N. (1984b). An expansion of the oscillatory source potential. *Appl. Ocean Res.* *6*, 116–117.

Newman, J. N. (1985). Algorithms for the free-surface Green function. *J. Engng. Math.* *19*, 57–67.

Newman, J. N. (1990). Numerical solutions of the water-wave dispersion relation. *Appl. Ocean Res.* *12*, 14–18.

Newman, J. N. (1992). Approximation of free-surface Green functions. In P. A. Martin and G. R. Wickham (Eds.), *Wave Asymptotics*, pp. 107–135. Cambridge University Press.

Newman, J. N. (1999). Radiation and diffraction analysis of the McIver toroid. *J. Engng. Math.* *35*, 135–147.

Noble, B. (1958). *Methods Based on the Wiener-Hopf Technique for the Solution of Partial Differential Equations.* Pergamon Press, London.

Noblesse, F. (1982). The Green function in the theory of radiation and diffraction of regular water waves by a body. *J. Engng. Math.* *16*, 137–169.

Ogilvie, T. F. (1963). First- and second-order forces on a cylinder submerged under a free surface. *J. Fluid Mech.* *16*, 451–472.

Ohkusu, M. (1974). Hydrodynamic forces on multiple cylinders in waves. In *Proc. Intl. Symposium on Dynamics of Marine Vehicles and Structures in Waves, London*, pp. 107–112.

O'Leary, M. (1985). Radiation and scattering of surface waves by a group of submerged, horizontal, circular cylinders. *Appl. Ocean Res. 7*, 51–57.

Omer, G. C. and H. H. Hall (1949). The scattering of a tsunami by a cylindrical island. *J. Seismological Soc. Am. 39*, 257–260.

Parsons, N. F. and P. A. Martin (1992). Scattering of water waves by submerged plates using hypersingular integral equations. *Appl. Ocean Res. 14*, 313–321.

Parsons, N. F. and P. A. Martin (1994). Scattering of water waves by submerged curved plates and by surface-piercing flat plates. *Appl. Ocean Res. 16*(3), 129–139.

Parsons, N. F. and P. A. Martin (1995). Trapping of water waves by submerged plates using hypersingular integral equations. *J. Fluid Mech. 284*, 359–375.

Penney, W. G. and A. T. Price (1952). The diffraction theory of sea waves and the shelter afforded by breakwaters. *Proc. Roy. Soc. Lond., A 244*, 236–253.

Petit, R. (Ed.) (1980). *Electromagnetic Theory of Gratings*, Volume 22 of *Topics in Current Physics*. Springer-Verlag, Berlin.

Porter, D. and P. G. Chamberlain (1997). Linear wave scattering by two-dimensional topography. In J. N. Hunt (Ed.), *Gravity Waves on Water of Finite Depth*, International Series on Advances in Fluid Mechanics. Computational Mechanics Publications, Southampton.

Porter, D. and K.-W. E. Chu (1986). The solution of two wave-diffraction problems. *J. Engng. Math. 20*, 63–72.

Porter, D. and D. S. G. Stirling (1990). *Integral Equations: A Practical Treatment from Spectral Theory to Applications*. Cambridge University Press.

Porter, D. and D. S. G. Stirling (1994). Finitely-generated solutions of certain integral equations. *Proc. Edinburgh Math. Soc. 37*, 325–345.

Porter, R. and D. V. Evans (1995). Complementary approximations to wave scattering by vertical barriers. *J. Fluid Mech. 294*, 155–180.

Porter, R. and D. V. Evans (1999). Rayleigh-Bloch surface waves along periodic gratings and their connection with trapped modes in waveguides. *J. Fluid Mech. 386*, 233–258.

Rayleigh, J. W. S. (1945). *The Theory of Sound. Volumes I and II.* Dover Publications, New York.

Row, R. V. (1955). Theoretical and experimental study of electromagnetic scattering by two identical conducting cylinders. *J. Appl. Phys. 26*, 666–675.

Rubin, H. (1954). The dock of finite extent. *Comm. Pure Appl. Maths 7*, 317–344.

Sarpkaya, T. and M. Isaacson (1981). *Mechanics of Wave Forces on Offshore Structures.* Van Nostrand Reinhold Company, New York.

Simon, M. J. (1982). Multiple scattering in arrays of axisymmetric wave energy devices, part 1: A matrix method using a plane-wave approximation. *J. Fluid Mech. 120*, 1–25.

Simon, M. J. (1985). The high frequency radiation of water waves by oscillating bodies. *Proc. Roy. Soc. Lond., A 401*, 89–115.

Simon, M. J. and F. Ursell (1984). Uniqueness in linearized two-dimensional water-wave problems. *J. Fluid Mech. 148*, 137–154.

Smallman, J. V. and D. Porter (1985). Wave diffraction by two inclined semi-infinite breakwaters. In *Proc. Intl. Conf. on Numerical and Hydraulic Modelling of Ports and Harbours*, Birmingham, U.K., pp. 269–278.

Smirnov, V. I. (1964). *A Course of Higher Mathematics. Vol IV. Integral Equations and Partial Differential Equations.* Pergamon Press, London.

Sparenberg, J. A. (1957). The finite dock. In *Proc. Symp. on the Behaviour of Ships in a Seaway*, Wageningen, pp. 717–728.

Spring, B. H. and P. L. Monkmeyer (1974). Interaction of plane waves with vertical cylinders. In *Proc. 14th Intl. Conf. on Coastal Engineering*, Copenhagen, pp. 1828–1845.

Spring, B. H. and P. L. Monkmeyer (1975). Interaction of plane waves with a row of cylinders. In *Proc. 3rd Conf. on Civil Eng. in Oceans, ASCE*, Newark, Delaware, pp. 979–998.

Srokosz, M. A. (1979). The submerged sphere as an absorber of wave power. *J. Fluid Mech. 95*, 717–741.

Srokosz, M. A. (1980). Some relations for bodies in a canal, with an application to wave-power absorption. *J. Fluid Mech. 99*, 145–162.

Srokosz, M. A. and D. V. Evans (1979). A theory for wave-power absorption by two independently oscillating bodies. *J. Fluid Mech. 90*, 337–362.

Stiassne, M. and G. Dagan (1973). Wave forces on a submerged vertical plate. *J. Engng. Math. 7*, 235–247.

Stoker, J. J. (1957). *Water Waves. The Mathematical Theory with Applications.* Interscience, New York.

Telste, J. G. and F. Noblesse (1986). Numerical evaluation of the Green function of water-wave radiation and diffraction. *J. Ship Res. 30*, 69–84.

Thomas, G. P. (1991). The diffraction of water waves by a circular cylinder in a channel. *Ocean Engng. 18*, 17–44.

Thomas, J. R. (1981). The absorption of wave energy by a three-dimensional submerged duct. *J. Fluid Mech. 104*, 189–215.

Thorne, R. (1953). Multipole expansions in the theory of surface waves. *Proc. Camb. Phil. Soc. 49*, 707–716.

Tuck, E. O. (1975). Matching problems involving flow through small holes. *Adv. Appl. Mech. 15*, 89–158.

Tung, C. C. (1979). Hydrodynamic forces on submerged vertical circular cylindrical tanks under ground excitation. *Appl. Ocean Res. 1*, 75–78.

Twersky, V. (1952). Multiple scattering of radiation by an arbitrary configuration of parallel cylinders. *J. Acoust. Soc. Am. 24*, 42–46.

Twersky, V. (1956). On the scattering of waves by an infinite grating. *IRE Trans. on Antennas and Propagation 4*, 330–345.

Twersky, V. (1961). Elementary function representation of Schlömilch series. *Arch. Rational Mech. Anal. 8*, 323–332.

Twersky, V. (1962). On scattering of waves by the infinite grating of circular cylinders. *IRE Trans. on Antennas and Propagation 10*, 737–765.

Ursell, F. (1947). The effect of a fixed vertical barrier on surface waves in deep water. *Proc. Camb. Phil. Soc. 43*, 374–382.

Ursell, F. (1948). On the waves due to the rolling of a ship. *Q. J. Mech. Appl. Math. 1*, 246–252.

Ursell, F. (1949). On the heaving motion of a circular cylinder on the surface of a fluid. *Q. J. Mech. Appl. Math. 2*, 218–231.

Ursell, F. (1950a). Surface waves on deep water in the presence of a submerged circular cylinder I. *Proc. Camb. Phil. Soc. 46*, 141–152.

Ursell, F. (1950b). Surface waves on deep water in the presence of a submerged circular cylinder II. *Proc. Camb. Phil. Soc. 46*, 153–158.

Ursell, F. (1951). Trapping modes in the theory of surface waves. *Proc. Camb. Phil. Soc. 47*, 347–358.

Ursell, F. (1962). Slender oscillating ships at zero forward speed. *J. Fluid Mech. 14*, 496–516.

Ursell, F. (1968). The expansion of water-wave potentials at great distances. *Proc. Camb. Phil. Soc. 64*, 811–826.

Ursell, F. (1973). On the exterior problems of acoustics. *Proc. Camb. Phil. Soc. 74*, 117–125.

Ursell, F. (1981). Irregular frequencies and the motion of floating bodies. *J. Fluid Mech. 105*, 143–156.

Ursell, F. (1999). On the wave motion near a submerged sphere between parallel walls: I. Multipole potentials. *Q. J. Mech. Appl. Math. 52*(4), 585–604.

Ursell, F. (2001). The local expansion of a source of oblique water waves in the free surface. *Wave Motion 33*, 109–116.

Utsunomiya, T. and R. Eatock Taylor (1999). Trapped modes around a row of circular cylinders in a channel. *J. Fluid Mech. 386*, 259–279.

VanBlaricum, G. F. and R. Mittra (1969). A modified residue calculus technique for solving a class of boundary value problems. Parts I and II. *IEEE Trans. Microwave Theory Tech. 17*(6), 302–319.

von Ignatowsky, W. (1914). Zur Theorie der Gitter. *Ann. Phys. 44*, 369.

Wang, S. (1981). Wave radiation due to oscillation of two parallel spaced cylinders. *Ocean Engng. 8*, 559–621.

Wang, S. (1986). Motions of a spherical submarine in waves. *Ocean Engng. 13*(3), 249–271.

Wang, S. and R. Wahab (1971). Heaving oscillations of twin cylinders in a free surface. *J. Ship Res. 15*, 33–48.

Watson, G. N. (1944). *A Treatise on the Theory of Bessel Functions* (2nd ed.). Cambridge University Press.

Wehausen, J. V. and E. V. Laitone (1960). Surface waves. *Handbuch der Physik 9*, 446–778.

Whitehead, E. A. N. (1951). The theory of parallel-plate media for microwave lenses. *Proc. IEE 98*(III), 133–140.

Wigley, N. M. (1964). Asymptotic expansions at a corner of solutions of mixed boundary value problems. *J. Math. Mech. 13*(4), 549–576.

Wilcox, C. H. (1984). *Scattering Theory for Diffraction Gratings*. Springer-Verlag, New York.

Williams, A. N. and W. W. Crull (1993). Wave diffraction by array of thin-screen breakwaters. *J. Wtrwy., Port, Coast., and Oc. Engrg., ASCE 119*, 606–617.

Williams, A. N. and M. K. Darwiche (1988). Three-dimensional wave scattering by elliptical breakwaters. *Ocean Engng. 15*(2), 103–118.

Williams, A. N. and M. K. Darwiche (1990). Wave radiation by truncated elliptical breakwaters. *J. Wtrwy., Port, Coast., and Oc. Engrg., ASCE 116*(1), 101–119.

Williams, M. H. (1982). Diffraction by a finite strip. *Q. J. Mech. Appl. Math. 35*, 103–124.

Wu, G. X. (1995). The interaction of water waves with a group of submerged spheres. *Appl. Ocean Res. 17*(3), 165–184.

Wu, G. X. (1998). Wave radiation and diffraction by a submerged sphere in a channel. *Q. J. Mech. Appl. Math. 51*(4), 647–666.

Wu, G. X. and R. Eatock Taylor (1987). The exciting force on a submerged spheroid in regular waves. *J. Fluid Mech. 182*, 411–426.

Wu, G. X. and R. Eatock Taylor (1989). On the radiation and diffraction of surface waves by submerged spheroids. *J. Ship Res. 33*(2), 84–92.

Yeung, R. W. (1981). Added mass and damping of a vertical cylinder in finite depth waters. *Appl. Ocean Res. 3*(3), 119–133.

Yeung, R. W. (1982). The transient heaving motion of floating cylinders. *J. Engng. Math. 16*, 97–119.

Yeung, R. W. and S. H. Sphaier (1989). Wave-interference effects on a truncated cylinder in a channel. *J. Engng. Math. 23*, 95–117.

Yilmaz, O. (1998). Hydrodynamic interactions of waves with group of truncated vertical cylinders. *J. Wtrwy., Port, Coast., and Oc. Engrg., ASCE 124*, 272–279.

Yilmaz, O. and A. Incecik (1998). Analytical solutions of the diffraction problem of a group of truncated vertical cylinders. *Ocean Engng. 25*, 385–394.

Yu, X. and A. T. Chwang (1993). Analysis of wave scattering by submerged circular disk. *J. Engrg. Mech. 119*(9), 1804–1817.

Yu, Y. S. and F. Ursell (1961). Surface waves generated by an oscillating circular cylinder on water of finite depth: theory and experiment. *J. Fluid Mech. 11*, 529–551.

Yue, D. K. P., H. S. Chen, and C. C. Mei (1978). A hybrid element method for diffraction of water waves by three dimensional bodies. *Intl. J. Numer. Meth. Eng. 12*, 245–266.

Záviška, F. (1913). Über die Beugung elektromagnetischer Wellen an parallelen, unendlich langen Kreiszylindern. *Ann. Phys. 40*, 1023–1056.

Index